Mathematics Is Not a Spectator Sport

George M. Phillips

Mathematics Is Not a Spectator Sport

With 68 Illustrations

 Springer

George M. Phillips
School of Mathematics and Statistics
University of St. Andrews
North Haugh
St. Andrews, Fife KY16 9SS
Scotland
gmp@st-andrews.ac.uk

Mathematics Subject Classification (2000): 00Axx, 11-xx, 20-xx, 51-xx

ISBN 978-1-4419-2061-4   e-ISBN 978-0-387-28697-6

Printed on acid-free paper.

9 8 7 6 5 4 3 2 1

springeronline.com

To my grandson Robert John Phillips

Preface

It is often said that mathematics and music go together, and that people with a special aptitude for mathematics often have similar gifts in music. Some music is very profound, and we find more in it at a second hearing. A similar point can be made about an understanding of mathematics.

In the world of music, there are *two* sets of people: active musicians who play musical instruments or sing, and the much larger set of passive musicians who listen to the sounds produced by members of the first set. However, in the world of mathematics, I contend that there is only *one* set of mathematicians: the active set. There are no such people as passive mathematicians. Of course, students attend mathematics lectures, and professional mathematicians take part in seminars in which a fellow mathematician discusses his or her research. But all who participate in such lectures and seminars, students and professional mathematicians alike, are active mathematicians, just as student musicians who attend a master class in their instrument are active musicians. In other words, there is nothing in the world of mathematics that corresponds to an audience in a concert hall, where the passive listen to the active. Happily, mathematicians are all *doers*, not spectators. *Doing* is much more fun than merely watching or listening, and I celebrate this in the title of this book.

Some very fine books have been written *about* mathematics. These give us the flavor of certain areas of mathematics, but they don't *make* us mathematicians. This book will introduce you to a range of topics in mathematics and will help you on your way to *becoming* a mathematician, even if you have only begun this journey. It is intended for senior students in high school and those who are beginning their study of mathematics at university level. There is a third category of readers that I have had very much in mind while writing this book. This is the large set of people who are not at school or university but are intellectually curious and active. If you are in this category, perhaps you have read that mathematics is more than just arithmetic and a little geometry, and want to know, what is it?

It seems to be generally accepted that the distribution of mathematical ability, like that of many other human characteristics, follows a bell-shaped curve (a Gaussian distribution), with comparatively few individuals at the two extremes. Relatively few individuals are either exceptionally poor or exceptionally good at mathematics. Thus many people have a much greater *potential* talent for mathematics than they think they have. As G. H. Hardy (1877–1947) writes in his fascinating book *A Mathematician's Apology* (see [11] in the References), "Most people are so frightened of the name of mathematics that they are ready, quite unaffectedly, to exaggerate their own mathematical stupidity." Many topics in mathematics are not immediately accessible; before we can understand them it is necessary to study some preliminary material. I have therefore chosen topics for this book that we can get into straight away.

Bertrand Russell (1872–1970) said that "A good notation has a subtlety and suggestiveness which at times make it almost seem like a live teacher." I agree with this wholeheartedly and have tried to explain carefully the notation that I use in this book. It is, in the main, notation that is widely used and is the common language of all mathematicians. I suppose that good notations have evolved by some kind of Darwinian natural selection process, and we need to be thoroughly familiar with mathematical notation in order to be able to construct or follow a mathematical argument.

With the possible exception of identical twins who have never been parted, and who have read all the same books and heard all the same things, we all have different backgrounds. So I can only try my best to guess, aided by my experience of writing and of lecturing, what I can best write in this book to get my message through to you. And on your part, you can try your best to understand me. I trust that our combined endeavors will be successful! But as you read this book, if some new notation puzzles you or if I fail to explain something to you adequately, I urge you to find someone who can help you lower the drawbridge and let you proceed into the "castle of mathematics." In writing the last sentence in more poetic language than is usual for me, I am mindful of the book *Drawbridge Up* (see reference [8]), in which the distinguished German poet and polymath Hans Magnus Enzensberger, who has a deep interest in mathematics, advances the thesis that mathematicians are often guilty of raising rather than lowering the drawbridge of understanding.

The earliest practitioners of the art of explaining mathematics are surely the Greeks, whose mathematics was developed during the millennium that began before the time of Pythagoras, in the sixth century BC. They introduced the concept of *proof* to mathematics. Thus, having identified the prime numbers, they did not just vaguely say that there are a lot of them; they *proved* that there is an infinite number of primes. Proofs are essential to mathematicians, and there are many proofs in this book. Greek mathematics was much concerned with geometry and number. At that time, numbers were regarded in a geometrical way, as lengths, or as ratios of lengths. This is very much the spirit of Chapter 1 of this book, which is concerned with a variety of results in number theory involving squares. It also includes material on Pythagoras's theorem, from the time of ancient Babylon and ancient China to a proof attributed to a nineteenth-century president of the United States. Chapter 1 concludes with a section on complex numbers.

Chapters 2 and 3 continue with the theme of numbers. Chapter 2 begins with a section on representing numbers in different bases, including base 2, which gives the binary system. It continues with a section on congruences, which involves algebraic ideas; a section on the fascinating arithmetic of continued fractions; and one on the Euclidean algorithm, whose origins are geometrical. Chapter 2 concludes with a section on the strange things that happen when we deal with infinite sets. This is based on the pioneering work of Georg Cantor (1845–1918) and introduces ideas that challenge our intuitive understanding of number.

Chapter 3 includes a study of the Fibonacci numbers, which are very simply defined, have many fascinating properties, and satisfy a large number of identities. They were introduced by Leonardo of Pisa eight centuries ago in his *Liber Abaci*, published in 1202. I describe how these numbers are related to the golden section number, involving ideas that go back to Pythagoras in the sixth century BC.

Chapter 4 is concerned with prime numbers. The Greek mathematician Eratosthenes devised a method for finding primes that has been applied in our own era, using the computer. It is sometimes said that mathematics is not an experimental subject. This is not true! Mathematicians often use the evidence of lots of examples to help form a conjecture, and this is an experimental approach. Having formed a conjecture about what might be true, the next task is to try to *prove* it. Thus, from near the beginning of the nineteenth century, the great mathematician C. F. Gauss (1777–1855) assembled lots of numerical evidence about how many primes there are in the first n numbers, where n is large. As I describe in Chapter 4, Gauss formed a conjecture about the *density* of primes, and a proof *was* found by others before the end of the nineteenth century. This chapter also describes special methods used in finding really large primes, and includes a discussion of the famous Riemann hypothesis. Although there is a massive amount of evidence to support the Riemann hypothesis, a proof has eluded mathematicians for about 150 years.

It is said that the early study of probability was prompted by questions put to mathematicians by gamblers who wanted to know the likelihood of various events that depend on chance. Probability theory is part of the content of Chapter 5, which is concerned with combinatorial mathematics: the number of ways of making choices. In this chapter I give a combinatorial interpretation of the Fibonacci numbers, providing a method for finding and verifying Fibonacci identities that supplements the more usual methods pursued in Chapter 3.

We return to geometry in Chapter 6, where we look at some of the geometrical constructions that were created by Greek mathematicians, and discuss some (like the trisection of an angle) that defeated them and were shown to be impossible two millennia later. This chapter includes a section on properties of the triangle, many of which were known to the ancient Greeks, and some which have been discovered since the nineteenth century. Chapter 6 also includes material on coordinate geometry, developed chiefly by Descartes. This makes it possible to solve certain *geometrical* problems by *algebraic* methods. You will already be familiar with the cube, but do you know that it is one of the five Platonic solids? I discuss these in the last section of Chapter 6, and also the thirteen Archimedean solids, which include one that inspired a widely used design of soccer balls.

By the nineteenth century some mathematicians had become aware that the study of certain symmetries could be developed using algebraic methods, a most exciting algebra in which ba could be different from ab. This led to the topic of group theory, which is discussed in Chapter 7. It is remarkable that such a glorious construct as group theory, in which mathematicians are still researching, is founded on four simple properties that can be stated in a few lines. I hope you will find this chapter rewarding, even if it proves to be more taxing than some of the other chapters. I have made it the last chapter for historical reasons and because it may require a more sustained effort on your part than the earlier chapters. This chapter ends with an algebraic interpretation of the Platonic solids. Here you will discover how even the humble regular tetrahedron (the Platonic solid that has four equilateral triangular faces) contains much more mathematics than one might suppose.

In writing a book of this length, with the aims stated above, I had to decide what to include and what to omit. As I have already said, I made a selection of varied topics that we can get into quickly, without lengthy preliminaries. These topics are largely independent, so that your understanding of a chapter is not seriously impaired by any (temporary!) difficulties with another chapter. However, I have drawn attention to connections between the various topics to show that they are indeed all constituent parts of one whole body of mathematics. One topic that I have omitted is the differential and integral calculus, which, although not especially difficult, does require some introductory material, such as a rigorous treatment of limits, making it inappropriate for a book such as this. Nonetheless, I have used the concept of limit a few times in the book, and rely on the reader's intuitive understanding of this concept. It is important in mathematics to appreciate when we have a rigorous proof, and when we have only an imprecise, intuitive understanding. I try at all times in this book to make it clear when I am *not* giving a proof.

One of the great charms of mathematics is its timeless quality. Results in mathematics that were obviously exciting and interesting to our predecessors in every era, from the time of ancient Babylon onwards, have the same fascination for us today. Mathematics seems to enjoy eternal youth. As a friend of Anthony says of Cleopatra in Shakespeare's *Anthony and Cleopatra*, "Age cannot wither her, nor custom stale her infinite variety." What is more, mathematics continues to develop, with no end in sight. Indeed, even within areas of the subject that have been studied for a long time, there are unsolved problems.

I hope you will enjoy *working* through this book, and that you will be happy at your work. You don't need to read it all at once. It is harder to read than a novel. Also, you don't need to read the chapters in the order in which they appear in the book. At a first reading, skim through the book and dip into it here and there. See what attracts you. If you get stuck, don't spend too long before finding someone who will help you. There are many avenues I haven't explored, even within the limited number of topics that I have discussed. Therefore, at some stage you may find yourself thinking that something or other is obviously true, and wonder why I haven't mentioned it. Then perhaps a little later, having learned something you didn't know before, *by your own efforts*, you will know that you are a mathematician. Good luck!

Acknowledgments

I owe a great deal to John Stillwell, who read the whole manuscript on behalf of my publisher and made very wise suggestions which I adopted wholeheartedly. I believe that the result is a much better book. It is also a pleasure to thank my friends Colin Campbell, Gracinda Gomes, and John Howie who read parts of my manuscript and kept me right on some things I didn't know enough about! I am grateful to my son Donald Phillips for drawing my attention to the work of Hans Magnus Enzensberger to which I refer in the Preface. I was very fortunate to have the support of David Kramer who, as my copyeditor for the third time, not only took the greatest care in scrutinizing every letter and punctuation mark of this text, but also made helpful comments on some mathematical points. Of course, any errors that remain are my sole responsibility. Finally, I wish to pay tribute to the fine work of those members of the staff of Springer, New York who have been involved with the production of this book, and especially thank my friends Mark Spencer and Ina Lindemann.

George M. Phillips
Crail, Scotland

Contents

1
Squares

I met a man once who told me that, far from believing in the square root of minus one, he didn't even believe in minus one.

E. C. Titchmarsh (1899–1963)

1.1 Square and Triangular Numbers

Everyone knows the whole numbers, also called the *positive integers* or the *natural numbers*,

$$1, 2, 3, 4, 5, 6, 7, 8, 9, 10, 11, 12, 13, 14, \ldots .$$

Young children often ask, "What is the biggest number?" and we have to answer, "There is no biggest number," since the sequence of positive integers goes on for ever. We say that the sequence of positive integers is *infinite*; this is what is meant by the row of dots after the number 14 above. Every positive integer is either *even* or *odd*. The even integers are those that are divisible by 2,

$$2, 4, 6, 8, 10, 12, 14, 16, 18, 20, 22, 24, 26, \ldots ,$$

and the odd integers are those that are *not* divisible by 2,

$$1, 3, 5, 7, 9, 11, 13, 15, 17, 19, 21, 23, 25, \ldots .$$

The number 2 is the smallest of the *prime numbers*, which we now define.

Definition 1.1.1 A prime number is a positive integer greater than 1 that is divisible only by 1 and itself. A number greater than 1 that is not prime is called *composite*.   ∎

The sequence of prime numbers begins 2, 3, 5, 7, 11, and I will have more to say about the primes later in this book.

Perhaps you recognize one or both of the following infinite sequences:

$$1,\ 4,\ 9,\ 16,\ 25,\ 36,\ 49,\ 64,\ 81,\ 100,\ 121,\ 144,\ \ldots\ , \qquad (1.1)$$

$$1,\ 3,\ 6,\ 10,\ 15,\ 21,\ 28,\ 36,\ 45,\ 55,\ 66,\ 78,\ 91,\ \ldots\ . \qquad (1.2)$$

The sequence (1.1) is the sequence of *squares*. Each square number can be depicted by an array of nodes arranged in a square formation, as in Figure 1.1, which shows the first six squares, 1, 4, 9, 16, 25, and 36.

FIGURE 1.1. The first six squares, 1, 4, 9, 16, 25, and 36.

The sequence (1.2) is the sequence of *triangular* numbers. The first triangular number is 1, and the next three are

$$1 + 2 = 3, \quad 1 + 2 + 3 = 6, \quad 1 + 2 + 3 + 4 = 10.$$

The first six members of this sequence are displayed as sets of nodes in triangular formation in Figure 1.2.

FIGURE 1.2. The first six triangular numbers, 1, 3, 6, 10, 15, and 21.

There is a famous story concerning triangular numbers and C. F. Gauss (1777–1855), who is generally regarded as one of the finest mathematicians of all time. When Gauss was in the early years of school his teacher decided to set the class an exercise that would keep them out of mischief for some time. The problem posed to the young children was to find the sum of the first hundred positive integers. To the teacher's astonishment, Carl Gauss almost immediately gave the correct answer, 5050. How did he do it? Gauss

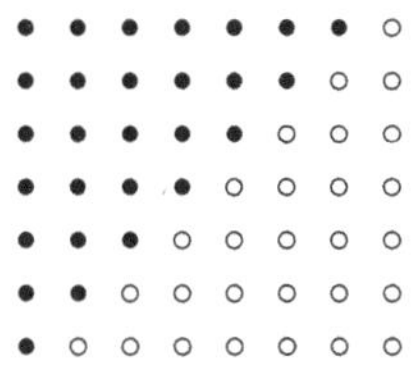

FIGURE 1.3. A geometrical interpretation of the equation $2T_n = n(n+1)$.

saw that the first and last numbers in the sum, 1 and 100, add up to 101, and also that the second and the second to last numbers, 2 and 99, add up to 101, and so on. It is easier to follow Gauss's argument if we write out the sum

$$T = 1 + 2 + 3 + \cdots + 98 + 99 + 100$$

and reverse the order of the numbers to give also

$$T = 100 + 99 + 98 + \cdots + 3 + 2 + 1.$$

I have called the sum T to remind us that it is a triangular number, and have used three dots in the latter two equations to denote the sum of the missing numbers, from 4 to 97. We can now add corresponding terms in the above two sums to give

$$2T = 101 + 101 + 101 + \cdots + 101 + 101 + 101.$$

Since there are a hundred terms in the latter sum, we see that

$$2T = 100 \times 101$$

and so

$$T = 50 \times 101 = 5050,$$

as Gauss obtained. Thus the 100th triangular number is 5050, and we can now follow Gauss's method to evaluate *any* triangular number. For if we write T_n to denote the nth triangular number, we have

$$T_n = 1 + 2 + 3 + \cdots + (n-2) + (n-1) + n,$$

and we can reverse the order of the terms in the above sum to give also

$$T_n = n + (n-1) + (n-2) + \cdots + 3 + 2 + 1.$$

Notice that if we replace n by 100 we obtain the two sums we had above for T. On adding corresponding terms in the above two sums for T_n, we obtain

$$2T_n = (n+1) + (n+1) + (n+1) + \cdots + (n+1) + (n+1) + (n+1),$$

and since there are n terms in the latter sum, we see that

$$2T_n = n(n+1),$$

meaning n multiplied by $n+1$. A geometrical interpretation of this equation is given in Figure 1.3, which shows that $2T_7 = 7 \times 8$. On dividing $2T_n$ by 2, we find that the triangular number T_n can be written in the form

$$T_n = \frac{1}{2}n(n+1). \tag{1.3}$$

As a check, we can evaluate this expression for some values of n, and we find, for example, that $T_3 = 6$, $T_5 = 15$, and $T_{100} = 5050$, in agreement with what we found above.

There is a special notation for handling sums. Suppose we have a sequence of numbers that begins with a_1, a_2, and a_3. We write the sum of the first n members of this sequence as

$$a_1 + a_2 + \cdots + a_n = \sum_{r=1}^{n} a_r. \tag{1.4}$$

The symbol $\sum$ is the uppercase letter *sigma*, the eighteenth letter of the classical Greek alphabet, and it stands for *sum*. We read the symbols on the right of the above equation as "sum a_r over all integer values of r from 1 to n." For example, we can express the nth triangular number as

$$T_n = \sum_{r=1}^{n} r.$$

There is a very simple connection between the triangular numbers and the squares, which follows immediately from Figure 1.4, where we see that every square is the sum of two *consecutive* triangular numbers. If we write S_n to denote n^2, the nth square, we have

$$S_n = T_{n-1} + T_n. \tag{1.5}$$

FIGURE 1.4. Every square is the sum of two consecutive triangular numbers.

This relation, which we have obtained by using a geometrical argument, can be verified algebraically. For we saw above that $T_n = \frac{1}{2}n(n+1)$, and so $T_{n-1} = \frac{1}{2}(n-1)n$. Thus we may write

$$T_{n-1} + T_n = \frac{1}{2}(n-1)n + \frac{1}{2}n(n+1) = \frac{1}{2}n\big((n-1) + (n+1)\big).$$

Since $(n-1) + (n+1) = 2n$, the above calculation shows that

$$T_{n-1} + T_n = n^2 = S_n,$$

which verifies the result displayed geometrically in Figure 1.4.

Having summed the first n positive integers, can we find the sum of the first n squares? This would settle the problem of determining how many oranges there are in a pyramid containing a square array of n^2 oranges on the bottom layer, an array of $(n-1)^2$ oranges on the second layer, and so on, with a single orange at the apex of the pyramid. Can you visualize how each layer of oranges nestles into the spaces between the oranges in the layer below? The similar problem of finding the sum of the first n triangular numbers, called the nth *tetrahedral number*, is equivalent to finding the number of oranges arranged in *triangular* layers of diminishing size. Let us imagine the oranges in each layer arranged in the form of an equilateral triangle (that is, with all three sides equal), rather than a right-angled triangle, as in Figure 1.2. The whole triangular configuration is called a triangular pyramid or a *tetrahedron*, which means "four faces" in Greek. Let us find the sum of the first n triangular numbers, and use that result to find the sum of the first n squares. We begin by writing

$$r(r+1)(r+2) - (r-1)r(r+1) = r(r+1)\big((r+2) - (r-1)\big).$$

If we divide the above equation throughout by 6 and simplify the right side, we find that

$$\frac{1}{6}r(r+1)(r+2) - \frac{1}{6}(r-1)r(r+1) = \frac{1}{2}r(r+1), \qquad (1.6)$$

and we observe that the number on the right of (1.6) is the rth triangular number. Our next move is to sum each number in (1.6) over r, from $r = 1$ to n, giving

$$\sum_{r=1}^{n} \frac{1}{6}r(r+1)(r+2) - \sum_{r=1}^{n} \frac{1}{6}(r-1)r(r+1) = \sum_{r=1}^{n} \frac{1}{2}r(r+1). \qquad (1.7)$$

At first sight, it may look as if we have made things worse, by expressing the sum of the first n triangular numbers in a more complicated way. But if we look more carefully at the left side of (1.7), we see that we are very close to the solution. For on putting $r = 1$ in the first sum on the left of (1.7), and $r = 2$ in the second sum, we find that these terms cancel. Likewise, the second term of the first sum cancels with the third term of the second sum, and so on. All that remains is the last term of the first sum and the first term of the second sum. Since the first term of the second sum is zero, we can express the nth tetrahedral number as

$$\sum_{r=1}^{n} T_r = \sum_{r=1}^{n} \frac{1}{2} r(r+1) = \frac{1}{6} n(n+1)(n+2). \tag{1.8}$$

As we saw in Figure 1.4,

$$S_r = T_{r-1} + T_r, \tag{1.9}$$

and we can use (1.9) and (1.8) to derive an expression for the sum of the first n squares. We have defined S_r and T_r for all positive integers r. We now define $T_0 = 0$, and then (1.9) will hold for all positive integers r. If we sum each number in (1.9) over r, from $r = 1$ to n, we obtain

$$\sum_{r=1}^{n} S_r = \sum_{r=1}^{n} T_{r-1} + \sum_{r=1}^{n} T_r.$$

Since

$$\sum_{r=1}^{n} T_{r-1} = \sum_{r=1}^{n-1} T_r,$$

we can use (1.8) to give

$$\sum_{r=1}^{n} S_r = \sum_{r=1}^{n-1} T_r + \sum_{r=1}^{n} T_r = \frac{1}{6}(n-1)n(n+1) + \frac{1}{6}n(n+1)(n+2).$$

We simplify the right side of the last equation, writing

$$\frac{1}{6}(n-1)n(n+1) + \frac{1}{6}n(n+1)(n+2) = \frac{1}{6}n(n+1)\big((n-1) + (n+2)\big),$$

and so obtain an expression for the sum of the first n squares,

$$\sum_{r=1}^{n} S_r = \frac{1}{6}n(n+1)(2n+1). \tag{1.10}$$

Above, we used a geometrical argument to show how a square can be expressed as the sum of two consecutive triangular numbers. Figure 1.5 shows another way of expressing a square as a sum. We begin with an $n \times n$ array of nodes and remove the L-shape of $2n - 1$ nodes consisting of those in the first column and the last row, leaving an $(n-1) \times (n-1)$ square. This geometrical observation corresponds to the algebraic equation

$$n^2 - (n-1)^2 = 2n - 1.$$

FIGURE 1.5. Every square is the sum of consecutive odd numbers.

We then remove an L-shape of $2n - 3$ nodes from the $(n - 1) \times (n - 1)$ square to leave an $(n - 2) \times (n - 2)$ square. After removing $n - 1$ L-shapes, only one node remains, showing that

$$n^2 = 1 + 3 + 5 + \ldots + (2n - 3) + (2n - 1) = \sum_{r=1}^{n}(2r - 1). \qquad (1.11)$$

This shows that n^2 is the sum of the first n odd numbers.

Mathematicians have been interested in square numbers at least since the development of Babylonian mathematics in the second millennium BC, between three and four thousand years ago. The number system used by the Babylonians was not very suitable for carrying out calculations, and perhaps this is why they devised the very clever method of multiplying two numbers that I will now describe.

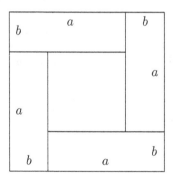

FIGURE 1.6. Geometrical interpretation of $(a + b)^2 = (a - b)^2 + 4ab$.

We write $a > b$ to mean a is greater than b, and write $a < b$ to mean a is less than b. We also write $a \geq b$ to mean a is greater than or equal to b, and then obviously $a \leq b$ means that a is less than or equal to b. Figure 1.6 consists of a large square of area $(a + b)^2$ that is subdivided into four identical rectangles of area ab and a small square of area $(a - b)^2$, where $a > b$. Thus we have

$$(a + b)^2 = (a - b)^2 + 4ab. \tag{1.12}$$

It follows from its geometrical derivation that this equation holds for all positive real values of a and b, with $a \geq b$. If $a = b$, the smaller square shrinks to zero, and the equation becomes $(a + a)^2 = 4a^2$. Can you see what happens to Figure 1.6 when $a = b$? We now make use of (1.12) by writing

$$x = \frac{1}{2}(a + b) \qquad \text{and} \qquad y = \frac{1}{2}(a - b). \tag{1.13}$$

Then it follows from (1.13) and (1.12) that

$$x^2 - y^2 = \frac{1}{4}(a + b)^2 - \frac{1}{4}(a - b)^2 = ab,$$

so that

$$ab = x^2 - y^2. \tag{1.14}$$

Equations (1.13) and (1.14) show how Babylonian mathematicians were able to multiply two numbers by using a table of squares. Suppose we have a table of squares of all positive integers from 1 to 10,000. Then, to multiply any two numbers $a > b$ of up to 4 decimal digits, we need only compute x and y from (1.13), look up their squares in the table, and then find ab from (1.14). We would need to modify this process if $a + b$ happens to be an odd number, because then $\frac{1}{2}(a+b)$ is not an integer. This happens only when one of the numbers a and b is even and the other is odd, and then $a + b$ and $a - b$ are both odd. (Problems 1.1.6 and 1.1.7 show how we can deal with this case.) If a and b both lie between 1 and 10,000, then since $a \times b = b \times a$, there are $T_{10,000} = 50,005,000$ possible multiplications. Thus the Babylonian method of multiplying two numbers would make a table of 10,000 squares equivalent to a gigantic multiplication table with more than fifty million entries.

Example 1.1.1 Let us evaluate 53×89 using the Babylonian method. Using (1.13) we compute

$$x = \frac{1}{2}(89 + 53) = 71 \qquad \text{and} \qquad y = \frac{1}{2}(89 - 53) = 18.$$

Then, using Table 1.1, we obtain from (1.14) that

$$89 \times 53 = 71^2 - 18^2 = 5041 - 324 = 4717. \qquad \blacksquare$$

In earlier eras, the need to perform multiplications arose in agrarian societies with the requirement to measure land areas, estimate amounts of agricultural products, and so on. There was little practical need for calculations other than simple additions, subtractions, multiplications, and

1	1	26	676	51	2601	76	5776
2	4	27	729	52	2704	77	5929
3	9	28	784	53	2809	78	6084
4	16	29	841	54	2916	79	6241
5	25	30	900	55	3025	80	6400
6	36	31	961	56	3136	81	6561
7	49	32	1024	57	3249	82	6724
8	64	33	1089	58	3364	83	6889
9	81	34	1156	59	3481	84	7056
10	100	35	1225	60	3600	85	7225
11	121	36	1296	61	3721	86	7396
12	144	37	1369	62	3844	87	7569
13	169	38	1444	63	3969	88	7744
14	196	39	1521	64	4096	89	7921
15	225	40	1600	65	4225	90	8100
16	256	41	1681	66	4356	91	8281
17	289	42	1764	67	4489	92	8464
18	324	43	1849	68	4624	93	8649
19	361	44	1936	69	4761	94	8836
20	400	45	2025	70	4900	95	9025
21	441	46	2116	71	5041	96	9216
22	484	47	2209	72	5184	97	9409
23	529	48	2304	73	5329	98	9604
24	576	49	2401	74	5476	99	9801
25	625	50	2500	75	5625	100	10000

TABLE 1.1. A table of squares.

divisions. However, even from the time of ancient Babylon, Greece, and China, mathematicians showed a sophisticated interest in the concept of number that was driven by sheer mathematical curiosity rather than commercial utility, and we will discuss some of these things later in this book. The growth of scientific knowledge, which greatly accelerated from the seventeenth century onwards, created a quantum leap in mankind's need to calculate, and encouraged the discovery of more efficient ways of carrying out arithmetical operations. In particular, logarithms were invented in the early years of the seventeenth century, and much effort went into the construction of logarithm tables. These were in constant use for carrying out multiplications and other calculations until the latter decades of the twentieth century, when the widespread availability of cheap electronic calculators made logarithm tables obsolete.

Although we can appreciate its ingenuity, and enjoy using it, the Babylonian method of multiplication using a table of squares is not necessary in our familiar decimal system, because calculations like those in Example 1.1.1 are easily carried out using "long multiplication." Let us work through such a calculation, for the sake of clarity. For example, we set out the evaluation of the product 53×89 in the form

$$
\begin{array}{r}
89 \\
\times 53 \\
\hline
267 \\
445 \\
\hline
4717
\end{array}
$$

In the above calculation we first multiply 89 by 3, the least-significant digit of 53, to give 267, and write it down. Then we multiply 89 by 5, the other digit of 53, to give 445, and write it down. We write 445 one place to the left of 267, because we are really multiplying 89 by 50 rather than 5. Finally, we add the last two numbers to give the result, which is 4717. To carry out such calculations, we need to know how to multiply two numbers a and b, where a and b are between 0 and 9, and add any carry digit. We also need to know how to write down the intermediate results, as in the above example, to take account of the decimal place.

The Roman numeral system is even more complicated than that of the Babylonians. Table 1.2 gives the Roman numerals that correspond to certain numbers in the system most familiar to us, which is called the Arabic or Hindu–Arabic system. Note the use of IV for 4, which is 5 minus 1, while VI denotes 6. Similarly, we write IX for 9, and XI for 11, XC for 90, and CX for 110, and so on. Table 1.2 gives us the basis for writing down all positive integers in Roman numerals up to a thousand, which is denoted by M. All other numbers are written down by combining those in the table in an obvious way. For example, we write XXIV for 24, CCCLV for 355, and MDCCCLXXXVIII for 1888. But if we contemplate how we would

Arabic	1	2	3	4	5	6	7	8	9
Roman	I	II	III	IV	V	VI	VII	VIII	IX

Arabic	10	20	30	40	50	60	70	80	90
Roman	X	XX	XXX	XL	L	LX	LXX	LXXX	XC

Arabic	100	200	300	400	500	600	700	800	900
Roman	C	CC	CCC	CD	D	DC	DCC	DCCC	CM

TABLE 1.2. The Arabic and Roman numeral systems.

multiply XLVII by LXXIII, for example, we realize that the Roman system is not convenient for doing arithmetic.

Before leaving the topic of multiplication, let us consider a method based on a table of triangular numbers instead of a table of squares. Consider Figure 1.7, which we can view as a rectangular array of 8×5 nodes together with a triangular array of $T_3 = 6$ nodes above it. The rectangular array of 40 nodes is split into a triangular array of $T_4 = 10$ nodes and a trapezoidal array of $30 = 4 + 5 + 6 + 7 + 8 = T_8 - T_3$ nodes, a trapezoidal figure being a quadrilateral (a four-sided figure) with only one pair of sides parallel.

In general, we can see that for any two positive integers $m \geq n$, we can split the rectangular array of nodes arranged in m columns and n rows into a triangular array of T_{n-1} nodes and a trapezoidal array containing $T_m - T_{m-n}$ nodes. Thus we have the remarkable result that

$$mn = T_m + T_{n-1} - T_{m-n}. \tag{1.15}$$

Notice how this equation agrees with (1.5) in the special case $m = n$.

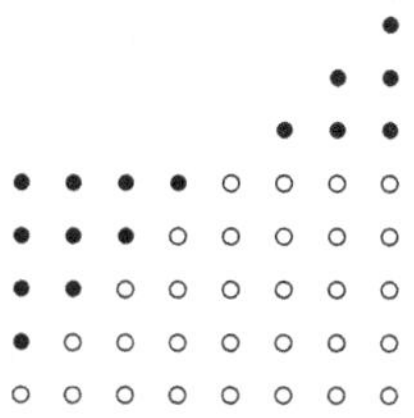

FIGURE 1.7. This diagram illustrates the equation $mn = T_m + T_{n-1} - T_{m-n}$, where $m = 8$ denotes the number of columns of nodes, and $n = 5$.

Example 1.1.2 Using (1.15) and Table 1.3 we find that

$$89 \times 53 = T_{89} + T_{52} - T_{36} = 4005 + 1378 - 666 = 4717$$

and

$$94 \times 38 = T_{94} + T_{37} - T_{56} = 4465 + 703 - 1596 = 3572. \quad \blacksquare$$

1	1	26	351	51	1326	76	2926
2	3	27	378	52	1378	77	3003
3	6	28	406	53	1431	78	3081
4	10	29	435	54	1485	79	3160
5	15	30	465	55	1540	80	3240
6	21	31	496	56	1596	81	3321
7	28	32	528	57	1653	82	3403
8	36	33	561	58	1711	83	3486
9	45	34	595	59	1770	84	3570
10	55	35	630	60	1830	85	3655
11	66	36	666	61	1891	86	3741
12	78	37	703	62	1953	87	3828
13	91	38	741	63	2016	88	3916
14	105	39	780	64	2080	89	4005
15	120	40	820	65	2145	90	4095
16	136	41	861	66	2211	91	4186
17	153	42	903	67	2278	92	4278
18	171	43	946	68	2346	93	4371
19	190	44	990	69	2415	94	4465
20	210	45	1035	70	2485	95	4560
21	231	46	1081	71	2556	96	4656
22	253	47	1128	72	2628	97	4753
23	276	48	1176	73	2701	98	4851
24	300	49	1225	74	2775	99	4950
25	325	50	1275	75	2850	100	5050

TABLE 1.3. A table of triangular numbers.

The number 1 is both a square and a triangular number, and an inspection of Tables 1.1 and 1.3 shows that 36 and 1225 also have this property. Such numbers correspond to solutions of the equation

$$n(n+1) = 2m^2, \tag{1.16}$$

as we see on dividing by 2. Equations for which we seek solutions in integers are called Diophantine equations, named after Diophantus of Alexandria, who lived in the third century AD, in the latter part of the glorious era of ancient Greek mathematics, which flourished for about a thousand years. Diophantine equations can be very difficult to solve. Even when we believe that a given Diophantine equation has *no* solution, this is often difficult to prove. The most famous, or notorious, Diophantine equation is that associated with Pierre de Fermat (1601–1665), to which (see (1.44)) we refer in Section 1.3. Happily, equation (1.16) is one for which we *can* find solutions. Let

$$\alpha = \left(1 + \sqrt{2}\right)^2 = 3 + 2\sqrt{2}, \tag{1.17}$$

where α is *alpha*, the first letter of the Greek alphabet. (The English word alphabet is derived from alpha and *beta*, the second letter of the Greek alphabet, which is written β.) We can show that equation (1.16) is satisfied by

$$m = m_k = \frac{1}{4\sqrt{2}} \left(\alpha^k - \alpha^{-k}\right), \tag{1.18}$$

where α^k means α multiplied by itself k times and $\alpha^{-k} = 1/\alpha^k$, and

$$n = n_k = \frac{1}{4} \left(\alpha^k + \alpha^{-k} - 2\right), \tag{1.19}$$

for every positive integer k. (See Problems 3.2.8 and 3.2.9 for a derivation of (1.18) and (1.19).) We will verify below that $m = m_k$ and $n = n_k$ are indeed positive integers, and that they satisfy (1.16). But first let us see why we might expect the presence of the quantity $\sqrt{2}$ in the above equations. We have from (1.16) that

$$2 = \frac{n(n+1)}{m^2} = \left(1 + \frac{1}{n}\right) \frac{n^2}{m^2}.$$

This shows that

$$\frac{2m^2}{n^2} = 1 + \frac{1}{n},$$

and we see that $2m^2/n^2$ tends to the limit 1 as n tends to infinity. Hence $n^2 \approx 2m^2$, where the symbol $\approx$ means "approximately equals," and so $n \approx \sqrt{2}\,m$ for large values of m and n.

Now let us replace n by $n+1$ in (1.19) to give

$$n+1 = \frac{1}{4} \left(\alpha^k + \alpha^{-k} + 2\right),$$

and so

$$n(n+1) = \frac{1}{16} \left(\alpha^k + \alpha^{-k} - 2 \right) \left(\alpha^k + \alpha^{-k} + 2 \right) = \frac{1}{16} \left(\left(\alpha^k + \alpha^{-k} \right)^2 - 4 \right).$$

This simplifies to give

$$n(n+1) = \frac{1}{16} \left(\alpha^{2k} + \alpha^{-2k} - 2 \right).$$

We find from (1.18) that

$$2m^2 = \frac{1}{16} \left(\alpha^{2k} + \alpha^{-2k} - 2 \right),$$

which shows that m and n defined by (1.18) and (1.19) do indeed satisfy $2m^2 = n(n+1)$, for all positive integers k. Finally, we need to show that these values of m and n are positive integers, and so justify our claim that there is an infinite number of squares that are also triangular numbers. One way of doing this is to begin by evaluating (1.18) and (1.19) with $k = 1$ and $k = 2$ to show that

$$m_1 = 1, \ m_2 = 6 \quad \text{and} \quad n_1 = 1, \ n_2 = 8. \tag{1.20}$$

In verifying (1.20) it is helpful to observe that

$$\left(3 + 2\sqrt{2} \right) \left(3 - 2\sqrt{2} \right) = 9 - 8 = 1,$$

and hence

$$\frac{1}{\alpha} = 3 - 2\sqrt{2}. \tag{1.21}$$

Then we show (see Problem 1.1.10) that

$$m_{k+1} = 6m_k - m_{k-1}, \quad k \geq 2, \tag{1.22}$$

and

$$n_{k+1} = 6n_k - n_{k-1} + 2, \quad k \geq 2. \tag{1.23}$$

We can compute as many values of m_k and n_k as we wish from (1.22) and (1.23), which are called *recurrence relations*, and we can see that m_k and n_k are indeed all positive integers. It follows from the recurrence relations that $m_3 = 35$ and $n_3 = 49$, and

$$35^2 = \frac{1}{2} 49 \cdot 50 = 1225,$$

which, as we found above, is the third number that is both a square and a triangular number. Although we have not proved it here, it can be shown that the only numbers that are both squares and triangular numbers are those obtained via the above recurrence relations. The next values of m_k and n_k are $m_4 = 204$, $n_4 = 288$, and $m_5 = 1189$, $n_5 = 1681$.

Problem 1.1.1 Show that the sum of the first n odd numbers can be expressed in the form

$$\sum_{r=1}^{n}(2r-1) = 2\sum_{r=1}^{n}r - \sum_{r=1}^{n}1 = 2T_n - n,$$

where T_n is the nth triangular number, and so verify that

$$\sum_{r=1}^{n}(2r-1) = n^2,$$

as we found using a geometrical argument.

Problem 1.1.2 Deduce from the equation

$$1+3+5+\cdots+(2n-1) = (1+2+3+\cdots+2n) - (2+4+6+\cdots+2n)$$

that

$$\sum_{r=1}^{n}(2r-1) = \sum_{r=1}^{2n}r - \sum_{r=1}^{n}2r = T_{2n} - 2T_n = n^2,$$

in agreement with the result obtained in Problem 1.1.1.

Problem 1.1.3 Imagine n^3 nodes arranged in a cube, just as we arranged n^2 nodes in a square in Figure 1.1. If we remove the nodes that lie on three faces of the cube that meet at a corner, we will be left with a cube of $(n-1)^3$ nodes. Verify that the number of nodes removed is $3n^2 - 3n + 1$, and deduce that

$$n^3 = \sum_{r=1}^{n}\left(3r^2 - 3r + 1\right).$$

Hence show that

$$\sum_{r=1}^{n}r^2 = \frac{1}{3}\left(n^3 - n\right) + \sum_{r=1}^{n}r = \frac{1}{3}\left(n^3 - n\right) + \frac{1}{2}n(n+1),$$

and so verify (1.10).

Problem 1.1.4 Verify that

$$n^2(n+1)^2 - (n-1)^2n^2 = 4n^3$$

and deduce that

$$\sum_{r=1}^{n}r^3 = \frac{1}{4}n^2(n+1)^2,$$

so that

$$1^3 + 2^3 + 3^3 + \cdots + n^3 = (1+2+3+\cdots+n)^2.$$

Problem 1.1.5 Let a and b be positive integers with $a + b$ odd. By considering the difference between $a + b$ and $a - b$, show that $a - b$ is also odd.

Problem 1.1.6 Let a and b be positive integers, with $a > b$ and $a+b$ odd. Then we may write

$$\frac{1}{2}(a + b) = d + \frac{1}{2},$$

where d is a positive integer. Show that

$$\frac{1}{2}(a - b) = d - b + \frac{1}{2},$$

and deduce from (1.12) that

$$ab = \left(d + \frac{1}{2}\right)^2 - \left(d - b + \frac{1}{2}\right)^2 = d^2 - (d - b)^2 + b.$$

Use this result and Table 1.1 to evaluate 86×57.

Problem 1.1.7 With the notation and conditions of Problem 1.1.6, write

$$\frac{1}{2}(a + b) = c - \frac{1}{2},$$

so that $c = d + 1$ is a positive integer, and deduce that

$$ab = c^2 - (c - b)^2 - b.$$

Problem 1.1.8 Verify (1.15) algebraically, using (1.3), and also show that

$$T_{m-1} + T_n - T_{m-n-1} = mn,$$

where $m > n$.

Problem 1.1.9 When he was only nineteen, Gauss proved that every positive integer can be expressed as the sum of three or fewer triangular numbers. Check that this property holds for the first 100 positive integers.

Problem 1.1.10 Deduce from (1.18) that

$$m_{k+1} + m_{k-1} = \frac{1}{4\sqrt{2}}\left(\left(\alpha + \frac{1}{\alpha}\right)\alpha^k - \left(\alpha + \frac{1}{\alpha}\right)\alpha^{-k}\right),$$

and use (1.17) and (1.21) to show that $\alpha + 1/\alpha = 6$, thus verifying the recurrence relation (1.22). Use the same method to verify (1.23).

1.2   Pythagoras's Theorem

One of the best known theorems in mathematics is Pythagoras's theorem, that in a right-angled triangle whose longest side has length c, and whose other two sides have lengths a and b, we have

$$a^2 + b^2 = c^2. \tag{1.24}$$

This is depicted in Figure 1.8. The longest side, which is opposite the right

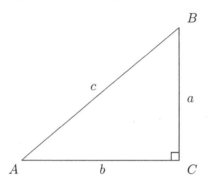

FIGURE 1.8. The sides of a right-angled triangle satisfy $a^2 + b^2 = c^2$.

angle, is called the *hypotenuse*. It is also true that if the sides of a triangle satisfy (1.24), then the angle opposite the longest side is a right angle. Although this theorem is named after the Greek mathematician Pythagoras, who lived in the sixth century BC, and the earliest known proof in the general case is due to the Greeks, the theorem was known much earlier to the Egyptians and even earlier to the Babylonians. The relation (1.24) holds for *all* right-angled triangles, for example the triangle whose sides are $a = b = 1$ and c, so that $c^2 = 2$. However, from the earliest times, there was an interest in finding right-angled triangles whose sides are positive integers. The simplest of these is the triangle whose sides are 3, 4, and 5. Eves [9] states that surveyors in ancient Egypt laid out right angles by constructing a 3, 4, 5 triangle using a rope divided into 12 equal parts by 11 knots, and that a proof that the 3, 4, 5 triangle is indeed right-angled was obtained in China, possibly as early as the second millennium BC.

To discuss this proof, we use Figure 1.9. It is clear by construction that there is a unique triangle, say T, that has sides of length 3 and 4 with a right angle contained between them. It also follows by construction that there is a unique triangle, say T', whose sides are of lengths 3, 4 and 5. We need to prove that the triangles T and T' are the same. We observe that two copies of the triangle T can be put together to form a rectangle with area 3×4, and so T has area 6. Figure 1.9 shows a quadrilateral $ABCD$ composed of a small square of unit length surrounded by four right-angled triangles that

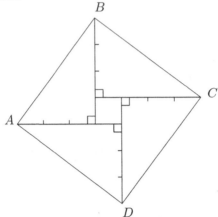

FIGURE 1.9. Chinese proof that the 3, 4, 5 triangle is right-angled.

are *congruent* to T. (Two or more geometrical figures that are identical are said to be congruent.) It follows that the area of the quadrilateral is $4 \times 6 + 1 = 25$. We next observe that because the four triangles are congruent, all four angles of the quadrilateral $ABCD$ are equal. Since, as we will prove in Problem 1.2.1, the sum of the angles of a quadrilateral is four right angles, $ABCD$ is a square. We then have

area of the square $ABCD = 25$,

and consequently AB has length 5. Thus the two triangles T and T' are congruent, which confirms that the 3, 4, 5 triangle is right-angled.

There are many proofs of Pythagoras's theorem. One proof, which is very easy to follow because of its geometric simplicity, is based on the two different dissections of a square shown in Figure 1.10. Both squares contain four right-angled triangles with sides a, b, and c. The square on the right, with area $(a + b)^2$, is dissected into the four triangles and two squares of areas a^2 and b^2. The square on the left, also of area $(a + b)^2$, is split into the four triangles and a quadrilateral whose four sides are all of length c. It is clear that the four angles of this quadrilateral are all equal, and so they

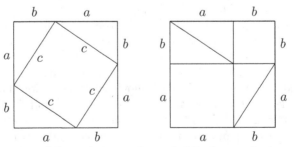

FIGURE 1.10. A pictorial proof of Pythagoras's theorem.

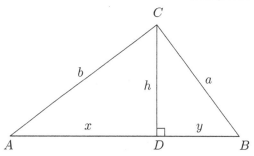

FIGURE 1.11. Proof of Pythagoras's theorem by similar triangles.

must all be right angles. Thus the quadrilateral is a square, with area c^2. We need only remove the four triangles from each of the two large squares of equal area to see that $a^2 + b^2 = c^2$, and this completes the proof.

We now give another simple proof of Pythagoras's theorem. This proof uses the notion of similar figures. Two figures drawn on the page are called *similar* if they have the same shape, apart from the *scale* of the figure. Thus, for example, all circles are similar, all squares are similar, and a 3, 4, 5 triangle is similar to a 9, 12, 15 triangle, but a 2×3 rectangle is not similar to a 3×5 rectangle.

Consider Figure 1.11, in which angle ACB is a right angle and CD is perpendicular to AB. Now, in any right-angled triangle, the sum of the two smaller angles is equal to a right angle. (We can see that this is true by putting two identical right-angled triangles together to make a rectangle.) Then we can deduce that the two smaller triangles in Figure 1.11, namely triangles ACD and CBD, have the same angles as the main triangle ABC, and so all three triangles are similar. In Figure 1.12, triangle ABC is the same as its namesake in Figure 1.11, and triangles $A'B'C$ and $A''B''C$ are obtained by cutting out triangles ACD and CBD from Figure 1.11, flipping them over, and pasting them into Figure 1.12. Thus the sum of the

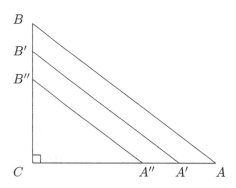

FIGURE 1.12. The three similar triangles in Figure 1.11.

areas of the two smaller triangles $A'B'C$ and $A''B''C$ is equal to the area of the largest triangle ABC. Now let

$$\frac{|BC|}{|AB|} = \frac{a}{c} = \lambda \quad \text{and} \quad \frac{|CA|}{|AB|} = \frac{b}{c} = \mu, \tag{1.25}$$

where $|BC|$ denotes the length of the line segment BC, and λ and μ (the Greek letters *lambda* and *mu*) are positive constants that apply to all triangles similar to triangle ABC. Thus $a = \lambda c$, $b = \mu c$, and we have

$$\Delta_c = \text{area of triangle } ABC = \frac{1}{2}ab = \frac{1}{2}\lambda\mu c^2, \tag{1.26}$$

where Δ is the uppercase Greek letter *delta*. Then, since triangles CBD and ABC are similar, we have

$$\frac{|BD|}{|BC|} = \frac{y}{a} = \lambda \quad \text{and} \quad \frac{|DC|}{|BC|} = \frac{h}{a} = \mu, \tag{1.27}$$

and it follows that

$$\Delta_a = \text{area of triangle } CBD = \frac{1}{2}yh = \frac{1}{2}\lambda\mu a^2. \tag{1.28}$$

We likewise derive from the similarity of triangles ACD and ABC that

$$\frac{|CD|}{|CA|} = \frac{h}{b} = \lambda \quad \text{and} \quad \frac{|DA|}{|CA|} = \frac{x}{b} = \mu, \tag{1.29}$$

so that

$$\Delta_b = \text{area of triangle } ACD = \frac{1}{2}hx = \frac{1}{2}\lambda\mu b^2. \tag{1.30}$$

Since $\Delta_a + \Delta_b = \Delta_c$, it follows from the above equations that

$$\frac{1}{2}\lambda\mu a^2 + \frac{1}{2}\lambda\mu b^2 = \frac{1}{2}\lambda\mu c^2,$$

and thus $a^2 + b^2 = c^2$. This completes our second proof of Pythagoras's theorem.

Among the very large number of proofs of Pythagoras's theorem, there is one that is attributed to James Garfield (1831–1881), who became the twentieth president of the United States in 1881. Garfield's proof depends on Figure 1.13, which consists of two right-angled triangles with sides a, b, and c, and half of a square of side c. Observe that we can put two copies of the trapezoidal shape $ABCD$ together to give the left-hand diagram in Figure 1.10, and so the quadrilateral $ABCD$ has area

$$\frac{1}{2}(a+b)^2 = \frac{1}{2}(a^2 + 2ab + b^2) = \frac{1}{2}(a^2 + b^2) + ab. \tag{1.31}$$

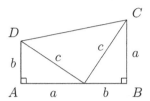

FIGURE 1.13. Garfield's proof of Pythagoras's theorem.

(The above expansion of $(a+b)^2$ can be verified by a geometrical argument, such as that given in Problem 1.2.3.) Continuing Garfield's proof, on adding the areas of the two triangles and the half square, we see that the area of the quadrilateral $ABCD$ can also be expressed as

$$\frac{1}{2}c^2 + ab. \tag{1.32}$$

We next subtract ab from each of the two expressions (1.31) and (1.32) for the area of the quadrilateral $ABCD$, to give $\frac{1}{2}(a^2+b^2) = \frac{1}{2}c^2$, and we need only multiply by 2 to complete the proof.

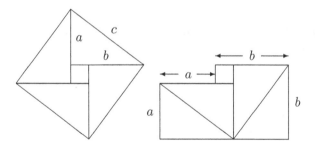

FIGURE 1.14. Chinese dissection method.

Many proofs of Pythagoras's theorem rely on dissecting a figure and rearranging it to give another shape with the same area, like the above proof based on Figure 1.10. Another such proof (see Figure 1.14) begins with four congruent right-angled triangles with sides a, b, and c placed around a square of side $b - a$, where a is the shortest side, to give a square of side c. This is rearranged to give a figure that can be viewed as two squares, one of side a and one of side b, nestling side by side. If $a = b$, the little square shrinks to zero.

Pythagoras's theorem appears as Proposition 47 in Book I of the famous set of books, the *Elements*, compiled by the Greek mathematician Euclid circa 300 BC. These books are a record of the finest achievements of Greek

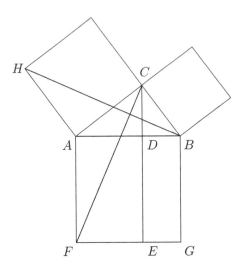

FIGURE 1.15. Euclid's diagram for Pythagoras's theorem.

mathematics up to that date. Euclid displays the right-angled triangle ABC (see Figure 1.15) and the squares constructed on each of its sides. The key step in Euclid's construction is to draw the line CE that splits the largest square into two rectangles. Euclid's proof depends on showing that each of these rectangles has the same area as the square that lies above it, so that the rectangle $ADEF$ has the same area as the square on the side AC, and the rectangle $DBGE$ has the same area as the square on the side BC. To do this, he first compares the triangles CAF and HAB. The *angles* CAF and HAB are equal, since each is a right angle plus the angle CAB. Since in each of the two triangles, the sides that enclose the equal angles CAF and HAB are also equal (that is, $|CA| = |HA|$ and $|AF| = |AB|$), it follows that the triangles CAF and HAB are congruent. Then, since the area of a triangle is half the length of its base times its height, the area of the triangle CAF is half the area of the rectangle $ADEF$, and also the area of triangle HAB is half the area of the square on AC. (See Problem 1.2.4.) Thus the the rectangle $ADEF$ has indeed the same area as the square on the side AC. Similarly, the rectangle $DBGE$ has the same area as the square on the side BC, and this completes Euclid's proof.

Problem 1.2.1 Put two congruent right-angled triangles together to form a rectangle, and hence show that the sum of the two smaller angles in the right-angled triangle add up to a right angle. Now, beginning with *any* triangle ABC, draw a perpendicular from a vertex to the opposite side, choosing the vertex so that the perpendicular lies within the triangle, thus dissecting the triangle ABC into two right-angled triangles. Hence show that the sum of the angles in triangle ABC is two right angles. Deduce that the sum of the angles of a quadrilateral is four right angles by splitting the quadrilateral into two triangles.

Problem 1.2.2 Deduce from (1.25), (1.27), and (1.29) that $h = ab/c$, $x = b^2/c$, and $y = a^2/c$. Show also that $\lambda^2 + \mu^2 = 1$.

Problem 1.2.3 By dissecting a square of side $a+b$ into four pieces, namely a square of side a, a square of side b, and two equal rectangles of area ab, show that
$$(a + b)^2 = a^2 + 2ab + b^2.$$

Problem 1.2.4 Verify that the area of a triangle is half the length of its base times its height, by considering the three types of triangle PQR depicted in the figure that follows.

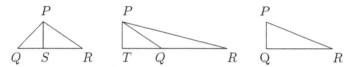

First show that this is true for the third figure, where the angle PQR is a right angle, since two copies of this triangle can be put together to form a rectangle. In the first figure, write $|QR| = |QS| + |SR|$ and express triangle PQR as the sum of two right-angled triangles, and in the second figure, write $|QR| = |TR| - |TQ|$ and express triangle PQR as the difference of two right-angled triangles.

Problem 1.2.5 Consider a triangle ABC. Let D denote the midpoint of BC, and let E denote the foot of the perpendicular from A to BC. (That is, the point where the line through A perpendicular to BC meets BC.) Consider the case in which E lies between B and D. Show that
$$|AB|^2 = |AE|^2 + |BE|^2 = |AD|^2 - |DE|^2 + |BE|^2,$$
so that
$$|AB|^2 = |AD|^2 + |BD|(|BE| - |DE|),$$
and similarly show that
$$|AC|^2 = |AD|^2 + |CD|(|EC| + |ED|).$$

By adding the latter two equations, show that
$$|AB|^2 + |AC|^2 = 2\left(|AD|^2 + |BD|^2\right). \tag{1.33}$$

Does (1.33) hold if E does not lie between B and D?

1.3   The Equation $a^2 + b^2 = c^2$

In this section we will write (a, b, c) to denote a triple of numbers. Let us consider the equation

$$a^2 + b^2 = c^2. \tag{1.34}$$

Obviously, if $c = \sqrt{a^2 + b^2}$, where a and b are any positive real numbers, the triples (a, b, c) and $(a, b, -c)$ are both solutions of (1.34). For example, $(1, 1, \sqrt{2})$ and $(1, 1, -\sqrt{2})$ satisfy (1.34). When we write $y = \sqrt{x}$, with x positive, y is defined as the *positive* number that satisfies $y^2 = x$. Thus $\sqrt{25} = 5$. It is more challenging to regard (1.34) as a Diophantine equation and seek solutions (a, b, c) in which a, b, and c are positive integers. Such a solution is called a *Pythagorean triple*, although much was known about such triples perhaps even a thousand years before the time of Pythagoras. We have already seen that $(3, 4, 5)$ is a Pythagorean triple. By trying some small values of a, b, and c, we find that $(5, 12, 13)$ is a Pythagorean triple, and you can find other solutions by such experimentation. However, we can do very much better than this because (1.34) is one of the rather rare Diophantine equations for which we can find *all* solutions. First we see that if (a, b, c) is a Pythagorean triple, then

$$(\lambda a)^2 + (\lambda b)^2 = \lambda^2 (a^2 + b^2) = \lambda^2 c^2 = (\lambda c)^2,$$

and so if λ is any positive integer, $(\lambda a, \lambda b, \lambda c)$ is also a Pythagorean triple. For example, $(6, 8, 10)$, $(9, 12, 15)$, and all other multiples of $(3, 4, 5)$ are Pythagorean triples. Thus we can concentrate on finding solutions (a, b, c) such that a, b, and c are positive integers with no common factor. This means that no two of a, b, and c have a common factor. For if any two of the numbers a, b, and c have a common factor, say m, we can deduce from (1.34) that the third number would also have m as a factor. A Pythagorean triple (a, b, c) such that a, b, and c have no common factor is called a *primitive* Pythagorean triple.

Obviously, a and b cannot both be even, for this means they would have the common factor 2. If $a = 2m + 1$, so that a is odd, we have

$$a^2 = (2m + 1)^2 = 4m^2 + 4m + 1$$

and thus a^2 has a remainder of 1 when we divide by 4. Thus if a and b are both odd, $c^2 = a^2 + b^2$ has a remainder of 2 when we divide by 4. Thus c must be even and so c^2 must be a multiple of 4. This gives a contradiction, since c^2 cannot have a remainder of 0 *and* 2 when we divide by 4. Thus a and b cannot both be odd; the only possibility is that one of the pair a and b is even, and the other is odd, and then c must be odd. Since it does not matter which is which between a and b, we will take a to be even and b to be odd. It follows that $c + b$ and $c - b$ are both even. Thus $\frac{1}{2}(c + b)$ and $\frac{1}{2}(c - b)$ are both positive integers, and they can have no common factor.

For if they had, we could show by adding and subtracting $\frac{1}{2}(c + b)$ and $\frac{1}{2}(c - b)$ that c and b would share this same common factor. Suppose now that p is an odd prime number that is a factor of a. Thus p^2 is a factor of a^2 and so must be is a factor of $\frac{1}{2}(c+b)$ or $\frac{1}{2}(c-b)$. Note that p cannot be a factor of both $\frac{1}{2}(c+b)$ and $\frac{1}{2}(c-b)$. For then the argument we used above concerning divisibility by 2 shows that p would be a factor of both b and c, and then (a, b, c) would not be a primitive triple. This crucial observation concerning division by a prime number plus a little further thought shows us that $\frac{1}{2}(c + b)$ and $\frac{1}{2}(c - b)$ must both be squares, say

$$\frac{1}{2}(c + b) = u^2 \quad \text{and} \quad \frac{1}{2}(c - b) = v^2,$$

where u and v are positive integers. On adding and subtracting the last two equations, we find that

$$c = u^2 + v^2 \quad \text{and} \quad b = u^2 - v^2. \tag{1.35}$$

It then follows from (1.34) and (1.12) that

$$a^2 = c^2 - b^2 = (u^2 + v^2)^2 - (u^2 - v^2)^2 = 4u^2v^2,$$

so that

$$a = 2uv. \tag{1.36}$$

The values of a, b, and c are said to be given in *parametric form* in terms of u and v in (1.35) and (1.36). We need to choose $u > v > 0$ such that a, b, and c are all positive, and further restrictions on u and v are required so that (1.35) and (1.36) yield a primitive solution (a, b, c). Obviously, u and v must have no common factor $d > 1$, for then a, b, and c would have d^2 as a common factor. Also, u and v cannot both be odd, for otherwise, a, b, and c would all have the common factor 2. Since any positive multiple of a primitive solution of (1.34) is also a solution, we have the following theorem.

Theorem 1.3.1 All solutions of the Diophantine equation

$$a^2 + b^2 = c^2$$

in positive integers are of the form

$$a = 2\lambda uv, \quad b = \lambda(u^2 - v^2), \quad c = \lambda(u^2 + v^2), \tag{1.37}$$

where λ, u, and v are positive integers such that $u > v$, u and v have no common factor greater than 1, and $u + v$ is odd. ∎

Given a solution of (a, b, c) of equation (1.34), we can determine unique values of λ, u, and v so that a, b, and c are given in the parametric form (1.37). The positive integer λ is just the greatest common divisor of a, b,

u	2	3	4	4	5	5	6	6	7	7	7
v	1	2	1	3	2	4	1	5	2	4	6
a	4	12	8	24	20	40	12	60	28	56	84
b	3	5	15	7	21	9	35	11	45	33	13
c	5	13	17	25	29	41	37	61	53	65	85

TABLE 1.4. The first few primitive Pythagorean triples (a, b, c).

and c. So it suffices to consider the case $\lambda = 1$. We then determine u and v uniquely as the positive numbers satisfying

$$u^2 = \frac{1}{2}(c + b) \qquad \text{and} \qquad v^2 = \frac{1}{2}(c - b),$$

where b is the smaller odd number, and c is the larger odd number. The first few primitive solutions of (1.34) are given in Table 1.4. We notice that each value of a in the table is a multiple of 4, and we can see that this holds for every a defined in (1.37), since either u or v is even.

Table 1.4 shows that the first Pythagorean triple $(3, 4, 5)$ is not the only one in which the sum of two consecutive squares is a square, since we also have $20^2 + 21^2 = 29^2$. It turns out that there is an infinite number of primitive Pythagorean triples that have this property. The key to the solution lies in the sequence (U_n) that we now define.

Consider the recurrence relation

$$U_{n+1} = 2U_n + U_{n-1}, \quad n \geq 1, \quad \text{with} \quad U_0 = 0, \, U_1 = 1. \tag{1.38}$$

This generates an infinite sequence whose next few values are $U_2 = 2$, $U_3 = 5$, $U_4 = 12$, $U_5 = 29$, and $U_6 = 70$. Let us now write

$$a = 2uv, \quad b = u^2 - v^2, \quad c = u^2 + v^2, \quad \text{where} \quad u = U_{n+1}, \, v = U_n, \tag{1.39}$$

so that a, b, and c are as defined in (1.35) and (1.36). Obviously, (a, b, c) is a Pythagorean triple. We will show that for any $n \geq 1$, a, b, and c have no common factor and $a - b = \pm 1$. Since

$$b - a = u^2 - v^2 - 2uv,$$

it will follow that $a - b = \pm 1$ if

$$W_n = U_{n+1}^2 - U_n^2 - 2U_n U_{n+1} = \pm 1. \tag{1.40}$$

It is easily verified that $W_0 = 1$ and $W_1 = -1$. If you compute the next few values of W_n, you will be encouraged by the fact that they continue to take the values 1 and -1 alternately. To *prove* that $W_n + W_{n+1} = 0$ for all values of n, we write

$$W_n + W_{n+1} = (U_{n+1}^2 - U_n^2 - 2U_n U_{n+1}) + (U_{n+2}^2 - U_{n+1}^2 - 2U_{n+1} U_{n+2}),$$

and so obtain

$$W_n + W_{n+1} = U_{n+2}^2 - U_n^2 - 2U_nU_{n+1} - 2U_{n+1}U_{n+2}. \tag{1.41}$$

Then, since

$$U_{n+2}^2 - U_n^2 = (U_{n+2} - U_n)(U_{n+2} + U_n),$$

we can simplify (1.41) to give

$$W_n + W_{n+1} = (U_{n+2} + U_n)(U_{n+2} - 2U_{n+1} - U_n) = 0. \tag{1.42}$$

Note that it follows from the recurrence relation (1.38) that the second factor on the right of (1.42) is zero.

Let us review what we know about the sequence (W_n). The initial values are $W_0 = 1$ and $W_1 = -1$, and we see from (1.42) that $W_{n+1} = -W_n$ for *all* n. Then, by the *principle of mathematical induction*, all members of the sequence (W_n) take the values $+1$ and -1 alternately, beginning with $W_0 = 1$. Mathematical induction has been likened to climbing a ladder. We need to get onto the first rung of the ladder, and when we are standing on *any* rung of the ladder, we need to be able to step up to the next rung. If these two conditions are met, we can climb the ladder.

Thus there is an infinite number of Pythagorean triples

$$(a, b, c) = (2uv, u^2 - v^2, u^2 + v^2), \quad \text{with} \quad u = U_{n+1}, \ v = U_n, \tag{1.43}$$

for which $a - b = \pm 1$. The Pythagorean triples (a, b, c) defined by (1.43) with values of n between 1 and 5 are

$$(4, 3, 5), \ (20, 21, 29), \ (120, 119, 169), \ (696, 697, 985), \ (4060, 4059, 5741).$$

The condition that $a - b = \pm 1$ implies that a and b have no common factor greater than 1, and thus all the Pythagorean triples defined by (1.43) are primitive.

Eves [9] writes that since the first half of the nineteenth century, about half a million clay tablets have been unearthed by archaeologists working in Mesopotamia, of which about 300 have shed light on ancient Babylonian mathematics. One of these tablets, housed at Columbia University, in New York, and known as Plimpton 322, is part of the collection named after G. A. Plimpton. From the style of the script used, it is thought to date from the period 1900 to 1600 BC. Plimpton 322 contains the b and c components of fifteen Pythagorean triples (a, b, c), of which all except two are primitive. The largest of the fifteen is the primitive triple that satisfies

$$13500^2 + 12709^2 = 18541^2.$$

Did the Babylonians discover such triples by a numerical search or through mathematical insight? I believe that these triples were found by people who knew what they were doing. If so, this was not blind, mechanical arithmetic but serious mathematics.

Pierre de Fermat famously asserted that the equation

$$a^n + b^n = c^n \qquad (1.44)$$

has no solutions in positive integers for $n > 2$, and even claimed he had a proof, although none was ever found. This came to be called Fermat's last theorem. A proof eluded mathematicians for more than 350 years until one was obtained by Andrew Wiles, and published in the *Annals of Mathematics* in 1995. Simon Singh [16] has given a fine nontechnical account of the story behind the solution of this problem. Even a proof that (1.44) has no solution in positive integers for exponent $n = 3$ is beyond the scope of this book. The usual proof given for the case $n = 4$ is very much simpler than that for $n = 3$. It relies on showing that the equation

$$x^4 + y^4 = z^2 \qquad (1.45)$$

has no solution in positive integers and thus (1.44) with $n = 4$ has no solution in positive integers. (If $x^4 + y^4$ cannot be a square it certainly cannot be a fourth power.) We begin by assuming that (1.45) *has* solutions, and let (x, y, z) denote the solution with the smallest value of z such that x, y, and z have no common factor greater than 1. We use the fact that (x^2, y^2, z) is a primitive Pythagorean triple. The proof is completed by deducing the existence of a *smaller* solution of (1.45), and this contradicts the above assumption that there is a solution of (1.45). This type of proof, which is called the *method of infinite descent*, was pioneered by Fermat. For a complete proof that the Diophantine equation (1.45) has no solution see Hardy and Wright [12] or Phillips [14].

Problem 1.3.1 If $u = 2m + 2n + 1$ and $v = 2m$, show that

$$u^2 - v^2 - 2uv - 1 = (u - v)^2 - 2v^2 - 1 = 4\left(n(n+1) - 2m^2\right),$$

and show that with $a = 2uv$ and $b = u^2 - v^2$, we have $b - a = 1$ for all values of m and n satisfying the Diophantine equation $n(n + 1) = 2m^2$. Which of the Pythagorean triples in Table 1.4 are of this form?

Problem 1.3.2 Given the primitive Pythagorean triple

$$(a, b, c) = (13500, 12709, 18541)$$

that was mentioned in the text, find values of u and v such that a, b, and c are given by (1.37) with $\lambda = 1$.

Problem 1.3.3 Verify that

$$(2n^2 + 2n)^2 + (2n + 1)^2 = (2n^2 + 2n + 1)^2$$

for all positive integers n and express the Pythagorean triple that satisfies the above equation in the form (1.37). Note that this gives Pythagorean triples (a, b, c) for which $c - a = 1$.

Problem 1.3.4 Show that

$$x = 2uv, \quad y = 2u^2 - v^2, \quad z = 2u^2 + v^2,$$

where u and v are positive integers such that $2u^2 > v^2$, is a solution of the Diophantine equation $2x^2 + y^2 = z^2$.

Problem 1.3.5 Find solutions of the Diophantine equation $3x^2 + y^2 = z^2$.

Problem 1.3.6 Let S_n denote the sum of the squares of the first n positive integers. Verify that $S_1 = 1$. *Assume* that for some integer $n \geq 1$,

$$S_n = \frac{1}{6}n(n + 1)(2n + 1) \tag{1.46}$$

and use the fact that $S_{n+1} = S_n + (n + 1)^2$ to deduce that

$$S_{n+1} = \frac{1}{6}(n + 1)(n + 2)(2n + 3).$$

Thus conclude by mathematical induction that (1.46) holds for all $n \geq 1$.

1.4   Sum of Two Squares

In this section we will discuss the question of which positive integers can be expressed as the sum of two squares. For example, we have $13 = 2^2 + 3^2$, but 3 cannot be expressed as a sum of two squares. I think it is appropriate to include this topic, because it fits in so well with the rest of the material in this chapter. However, I will not prove all of the results that I state in this section, because that would take us deeper into the theory of numbers than seems appropriate in this book. The reader who wishes to gain a fuller understanding of the material will find the omitted proofs in Hardy and Wright [12], or in Phillips [14]. We begin by stating without proof that every positive integer n may be written uniquely in the form

$$n = p_1^{\alpha_1} p_2^{\alpha_2} p_3^{\alpha_3} \cdots p_N^{\alpha_N} \tag{1.47}$$

for some choice of N, where $p_1 = 2$, $p_2 = 3$, $p_3 = 5$, and so on, are the prime numbers, each exponent $\alpha_1, \alpha_2, \alpha_3, \ldots \alpha_{N-1}$ is a nonnegative integer, and the last exponent, α_N, is positive. Thus if an exponent α_j is zero, it means that p_j is not present in the above representation of n. Given any n, we can derive the unique factorization (1.47) by testing whether n is divisible by $p_1 = 2$. We divide by 2 as many times as we can. Then, beginning with the quotient that remains after repeated division by 2, we divide this by $p_2 = 3$ as many times as we can. We repeat this process with successive primes until we obtain the quotient 1. For example, we find that

$$56852 = 2^2 \times 61 \times 233 \quad \text{and} \quad 550368 = 2^5 \times 3^3 \times 7^2 \times 13.$$

Beginning with the representation of n as a product of powers of primes, as in (1.47), we can derive a second expression for n that is also unique, writing

$$n = \lambda^2 n_1, \qquad \text{with} \qquad n_1 = p_1^{\beta_1} p_2^{\beta_2} p_3^{\beta_3} \cdots p_N^{\beta_N},$$

where $\lambda \geq 1$ is a positive integer and each exponent β_j is either 0 or 1. We then say that the above number n_1 is *square-free*, since it is not divisible by any square greater than 1. Observe that a square-free number is simply a product of distinct primes. For example, we have

$$n = 550368 = 2^5 3^3 7^2 13 = (2^4 3^2 7^2) \times 2 \times 3 \times 13 = (2^2 \times 3 \times 7)^2 \times 2 \times 3 \times 13,$$

so that $\lambda = 2^2 \times 3 \times 7 = 84$, and the number $n_1 = 2 \times 3 \times 13 = 78$ is square-free. If we can express n_1 as the sum of two squares, say

$$n_1 = a^2 + b^2,$$

then we can write n as the sum of two squares, since

$$n = \lambda^2 n_1 = (\lambda a)^2 + (\lambda b)^2.$$

This greatly simplifies our task of finding all positive integers that can be expressed as the sum of two squares. For we have reduced our original problem to that of finding which square-free numbers can be expressed as the sum of two squares.

Apart from 2, the first prime, every prime is odd and must have one of the forms $4m + 1$ or $4m + 3$. Consider the following two statements concerning these two classes of odd primes.

1. Every prime number of the form $4m + 1$ can be expressed as the sum of two squares, and this can be done uniquely.

2. No prime number of the form $4m + 3$ can be expressed as the sum of two squares.

Both statements are true. The second statement is easily verified, as follows. We saw in the last section that a^2 is divisible by 4 if a is even, and leaves a remainder of 1 on division by 4 if a is odd. Thus $a^2 + b^2$ has a remainder of either 0, 1, or 2 on division by 4. It follows that if n is of the form $4m + 3$, it cannot be expressed as the sum of two squares. The first of the above statements is very much harder to prove than the second. One proof (see Hardy and Wright [12] or Phillips [14]) relies on the use of complex numbers, which we will meet in the next section.

We further state without proof that if a square-free n has any prime factor of the form $4m + 3$, then n *cannot* be expressed as the sum of two squares. (See Hardy and Wright [12].) Finally, we make use of the identity

$$(a^2 + b^2)(c^2 + d^2) = (ac - bd)^2 + (ad + bc)^2. \tag{1.48}$$

This identity was used implicitly by Diophantus of Alexandria in his book *Arithmetica*, which was published in the third century AD. It was quoted by Leonardo of Pisa (circa 1175–1220), who is also known as Fibonacci, in his book *Liber Abaci*, published in 1202. By repeatedly applying (1.48), we can deduce that any square-free n whose prime factorization may or may not contain 2 and otherwise consists only of odd primes of the form $4m + 1$ is expressible as the sum of two squares. Furthermore, it follows from what has been said above that these are the only square-free numbers that are expressible as the sum of two squares.

Observe that (1.48) reduces to

$$2(a^2 + b^2) = (a - b)^2 + (a + b)^2 \tag{1.49}$$

when we put $c = d = 1$. Note also that if $a \neq b$ (meaning a is not equal to b) and $c \neq d$, we can interchange c and d in (1.48) and so obtain two different expressions of $(a^2 + b^2)(c^2 + d^2)$ as the sum of two squares.

Example 1.4.1 We have

$$2 = 1^2 + 1^2, \qquad 13 = 3^2 + 2^2, \qquad 41 = 5^2 + 4^2,$$

and we have from (1.49) that $26 = 2(3^2 + 2^2) = 1^2 + 5^2$. We can now use (1.48) to give

$$26 \times 41 = (1^2 + 5^2)(5^2 + 4^2) = 15^2 + 29^2$$

and also

$$26 \times 41 = (1^2 + 5^2)(4^2 + 5^2) = 21^2 + 25^2.$$

Thus we can write 1066 as the sum of two squares in two ways,

$$1066 = 15^2 + 29^2 = 21^2 + 25^2. \qquad \blacksquare$$

Problem 1.4.1 Show that 1776 cannot be expressed as the sum of two squares.

Problem 1.4.2 Express 1314 as the sum of two squares.

Problem 1.4.3 C. F. Gauss was born in Braunschweig in 1777 and died in Göttingen in 1855. Express one of these numbers as the sum of two squares, and show that the other number cannot be expressed in this form.

Problem 1.4.4 Let n be an odd integer such that

$$2n^2 = u^2 + v^2,$$

where u and v are positive integers and $u > v$. Show that u and v must both be odd and that consequently, $\frac{1}{2}(u+v)$ and $\frac{1}{2}(u-v)$ are both positive integers. Deduce that n may be expressed as the sum of two squares, in the form

$$n^2 = \left(\frac{1}{2}(u+v)\right)^2 + \left(\frac{1}{2}(u-v)\right)^2.$$

Problem 1.4.5 Find the smallest integer that can be expressed as the sum of two distinct squares in two different ways.

1.5   Complex Numbers

Consider the *quadratic* equation

$$az^2 + bz + c = 0, \tag{1.50}$$

where a, b, and c are any real numbers, with a nonzero. It is called a quadratic equation because it involves the square (Latin *quadrum*) of the unknown quantity z. We will find all values of z that satisfy it. These are called *solutions* of the equation. We begin by dividing the quantities on each side of the equation by the nonzero number a, giving

$$z^2 + \left(\frac{b}{a}\right)z + \frac{c}{a} = 0. \tag{1.51}$$

It should be clear that the two equations (1.50) and (1.51) have the same solutions. If we subtract c/a from both sides of (1.51) we obtain

$$z^2 + \left(\frac{b}{a}\right)z = -\frac{c}{a}, \tag{1.52}$$

which also has the same solutions as the two earlier equations. We now "complete the square," adding a suitable constant to each side of (1.52) so that we can write the left side in the form $(z + \alpha)^2$, where α is real. Since

$$\left(z + \frac{b}{2a}\right)^2 = z^2 + \left(\frac{b}{a}\right)z + \frac{b^2}{4a^2},$$

the quantity we need to add to both sides of (1.52) is $b^2/4a^2$, giving

$$\left(z + \frac{b}{2a}\right)^2 = -\frac{c}{a} + \frac{b^2}{4a^2},$$

which can be written in the form

$$\left(z + \frac{b}{2a}\right)^2 = \frac{b^2 - 4ac}{4a^2}. \tag{1.53}$$

This quadratic equation has the same solutions as each of the equations (1.50), (1.51), and (1.52). We now see that there are three possibilities, depending on whether $b^2 - 4ac$ is zero, positive, or negative.

1. If $b^2 - 4ac$ is zero, it follows from (1.53) that equation (1.50) has only one solution, namely

$$z = -\frac{b}{2a}. \tag{1.54}$$

2. If $b^2 - 4ac$ is positive, we see from (1.53) that

$$z + \frac{b}{2a} = +\frac{\sqrt{b^2 - 4ac}}{2a} \quad \text{or} \quad z + \frac{b}{2a} = -\frac{\sqrt{b^2 - 4ac}}{2a},$$

where $\sqrt{b^2 - 4ac}$ is the positive number whose square is $b^2 - 4ac$. In this case the quadratic equation (1.50) has the two solutions

$$z = -\frac{b}{2a} + \frac{\sqrt{b^2 - 4ac}}{2a} \quad \text{and} \quad z = -\frac{b}{2a} - \frac{\sqrt{b^2 - 4ac}}{2a}.$$

3. If $b^2 - 4ac$ is *negative*, we have the most interesting case! It was said before the invention of complex numbers that (1.53), and consequently (1.50), has *no solutions* when $b^2 - 4ac$ is negative, because a negative number has no square root. However, as we will see below, we *now* say that (1.53) has no solutions in real numbers, but has two solutions that are complex numbers.

Consider the quadratic equation $z^2 + 1 = 0$, which is equivalent to

$$z^2 = -1. \tag{1.55}$$

Since the square of any real number is nonnegative, it is clear that this equation has no solution in real numbers. Let us assume that there is some other system of "numbers" in which there is a solution of (1.55). Let us denote such a solution by i, so that $i^2 = -1$. We will also assume that this number system behaves algebraically like the real numbers so that, for example, any number α in this system has the property that $\alpha^2 = (-\alpha)^2$. This entails that (1.55) has the two solutions $z = i$ and $z = -i$. We then find that

$$(z - i)(z + i) = z(z + i) - i(z + i) = z^2 + iz - iz - i^2,$$

and since $i^2 = -1$, we obtain

$$(z - i)(z + i) = z^2 + 1.$$

Thus we have expressed $z^2 + 1$ as a product of the two factors $z - i$ and $z + i$. Then $z^2 + 1 = 0$ is equivalent to

$$(z - i)(z + i) = 0,$$

and so $z - i = 0$ or $z + i = 0$. Although this tells us only what we already know, that $z = i$ or $z = -i$, it gives us some confidence in the algebra of this system.

We can now pursue the solution of (1.53) for the case $b^2 - 4ac < 0$. For then

$$\left(z + \frac{b}{2a} \right)^2 = \frac{4ac - b^2}{4a^2} \cdot (-1),$$

and so

$$z + \frac{b}{2a} = \frac{\sqrt{4ac - b^2}}{2a} i \quad \text{or} \quad z + \frac{b}{2a} = -\frac{\sqrt{4ac - b^2}}{2a} i,$$

where $\sqrt{4ac - b^2}$ is the positive number whose square is the positive number $4ac - b^2$. Thus, when $b^2 - 4ac < 0$, the equation (1.50) has the two solutions

$$z = -\frac{b}{2a} + \frac{\sqrt{4ac - b^2}}{2a} i \quad \text{and} \quad z = -\frac{b}{2a} - \frac{\sqrt{4ac - b^2}}{2a} i. \qquad (1.56)$$

These have the form $z = x + yi$ and $z = x - yi$, where the two numbers

$$x = -\frac{b}{2a} \quad \text{and} \quad y = \frac{\sqrt{4ac - b^2}}{2a}$$

are both real and $i^2 = -1$. Any number of the form $x + yi$, with x and y real, is called a *complex number*.

Definition 1.5.1 Given the complex number $z = x + yi$, we say that x is the *real* part of z and y is the *imaginary* part. We write the real and imaginary parts of z as

$$x = \mathbf{Re}(z) \quad \text{and} \quad y = \mathbf{Im}(z). \qquad \blacksquare \qquad (1.57)$$

Remark 1.5.1 The terms "real part" and "imaginary part" are a little misleading, since mathematicians would agree that the imaginary part of a complex number is no more "imaginary," in the everyday sense of the word, than the real part. One might also say that a real number is no more or less "real," in the everyday sense, than any other kind of number. $\blacksquare$

We can evaluate sums and products of complex numbers in an obvious way. Thus, if $z_1 = a + bi$ and $z_2 = c + di$, their sum is

$$z_1 + z_2 = (a + c) + (b + d)i, \tag{1.58}$$

and their product is

$$z_1 z_2 = (a + bi)(c + di) = a(c + di) + bi(c + di) = ac + adi + bci + bdi^2.$$

Since $i^2 = -1$, we find that

$$z_1 z_2 = (a + bi)(c + di) = (ac - bd) + (ad + bc)i. \tag{1.59}$$

Note from (1.58) and (1.59) that the sum and product of two complex numbers are both complex numbers. If we put $a = c = x$ and $b = -d = y$ in (1.59), we obtain, as a special case,

$$(x + yi)(x - yi) = x^2 + y^2. \tag{1.60}$$

Definition 1.5.2 If $z = x + yi$, we denote $x - yi$ by $\bar{z}$ and call $\bar{z}$ the *conjugate* of z. It follows from this definition that the conjugate of $\bar{z}$ is z, and we refer to z and $\bar{z}$ as a complex conjugate pair.   ■

It follows from (1.56) that when the solutions of the quadratic equation (1.50) are complex, the two solutions are a complex conjugate pair.

We see from Definitions 1.5.1 and 1.5.2 that

$$z + \bar{z} = 2\,\mathbf{Re}(z) \quad \text{and} \quad z - \bar{z} = 2i\,\mathbf{Im}(z). \tag{1.61}$$

Definition 1.5.3 If $z = x + yi$, we write $|z|$ to denote $\sqrt{x^2 + y^2}$ and call $|z|$ the *modulus* of the complex number z.   ■

Since $|x| \leq \sqrt{x^2 + y^2}$ and $|y| \leq \sqrt{x^2 + y^2}$, we see from Definitions 1.5.1 and 1.5.3 that

$$|\mathbf{Re}(z)| \leq |z| \quad \text{and} \quad |\mathbf{Im}(z)| \leq |z|. \tag{1.62}$$

It follows from Definitions 1.5.2 and 1.5.3 that $|\bar{z}| = |z|$, and it is then clear from (1.60) that for any complex number z,

$$z\bar{z} = |z|^2. \tag{1.63}$$

With $z_1 = a + bi$ and $z_2 = c + di$, we see from (1.59) that

$$|z_1 z_2|^2 = (ac - bd)^2 + (ad + bc)^2,$$

and we deduce from the identity (1.48) that $|z_1 z_2|^2 = |z_1|^2 |z_2|^2$, and thus

$$|z_1 z_2| = |z_1|\,|z_2|. \tag{1.64}$$

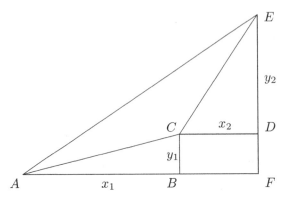

FIGURE 1.16. If $|AB| = x_1$, $|BC| = y_1$, $|CD| = x_2$, and $|DE| = y_2$, then $|AF| = x_1 + x_2$ and $|FE| = y_1 + y_2$. With $z_1 = x_1 + y_1 i$ and $z_2 = x_2 + y_2 i$, we have $|AC| = |z_1|$, $|CE| = |z_2|$, and $|AE| = |z_1 + z_2|$.

It is remarkable that the wonderful identity (1.48) was known in number theory fifteen hundred years before its application to complex numbers was appreciated. We will meet a generalization of (1.48) in Problem 7.1.9.

The modulus of a complex number behaves like a *length*. Indeed, if x and y are positive, the modulus of $z = x + yi$ is the length of the hypotenuse of a right-angled triangle whose shorter sides are x and y. This leads to the *triangle inequality*,

$$|z_1 + z_2| \leq |z_1| + |z_2|, \tag{1.65}$$

as demonstrated geometrically by the inequality $|AC| + |CE| < |AE|$ in Figure 1.16. This would be an equality if A, C, and E were in a straight line. The triangle inequality can also be justified algebraically. Let us assume that z_1 and z_2 are both nonzero, for otherwise, (1.65) is trivial. We first verify that the conjugate of $z_1 + z_2$ is $\bar{z}_1 + \bar{z}_2$ and that $z_1 \bar{z}_2$ and $\bar{z}_1 z_2$ are conjugates. Then we have from (1.63) that

$$|z_1 + z_2|^2 = (z_1 + z_2)(\bar{z}_1 + \bar{z}_2) = z_1 \bar{z}_1 + z_1 \bar{z}_2 + \bar{z}_1 z_2 + z_2 \bar{z}_2. \tag{1.66}$$

Since $z_1 \bar{z}_2$ and $\bar{z}_1 z_2$ are conjugates, we see from (1.61) that

$$z_1 \bar{z}_2 + \bar{z}_1 z_2 = 2 \operatorname{Re}(z_1 \bar{z}_2).$$

If we now use the latter equality and (1.63) in (1.66), we find that

$$|z_1 + z_2|^2 = |z_1|^2 + 2 \operatorname{Re}(z_1 z_2) + |z_2|^2.$$

Then, on using the first inequality in (1.62), we obtain

$$|z_1 + z_2|^2 \leq |z_1|^2 + 2 |z_1 \bar{z}_2| + |z_2|^2.$$

Finally, we use (1.64) and the property that $|\bar{z}| = |z|$ to give

$$|z_1 + z_2|^2 \leq |z_1|^2 + 2 |z_1| \, |z_2| + |z_2|^2 = (|z_1| + |z_2|)^2,$$

and hence (1.65) holds. This will be an equality if

$$\mathbf{Re}\,(z_1 \bar{z}_2) = |z_1 \bar{z}_2|,$$

which will hold if and only if

$$\mathbf{Re}\,(z_1 \bar{z}_2) \geq 0 \quad \text{and} \quad \mathbf{Im}\,(z_1 \bar{z}_2) = 0. \tag{1.67}$$

Let us write $z_1 = x_1 + y_1 i$ and $z_2 = x_2 + y_2 i$. Then

$$\mathbf{Re}\,(z_1 \bar{z}_2) = x_1 x_2 + y_1 y_2 \quad \text{and} \quad \mathbf{Im}\,(z_1 \bar{z}_2) = x_2 y_1 - x_1 y_2.$$

We can see that if z_2 is a positive multiple of z_1, so that

$$x_2 = \lambda x_1 \quad \text{and} \quad y_2 = \lambda y_1, \quad \text{with } \lambda > 0, \tag{1.68}$$

then *both* conditions in (1.67) will hold, and thus (1.65) will be an equality. The conditions in (1.68) are equivalent to the geometrical condition that the three points A, C, and E in Figure 1.16 lie in a straight line, with C lying between A and E. See also Problem 1.5.5.

We can generalize the quadratic equation, defined by (1.50), to give a *polynomial* equation of degree n,

$$a_0 z^n + a_1 z^{n-1} + \cdots + a_{n-1} z + a_n = 0, \tag{1.69}$$

where the coefficients $a_0, a_1, \ldots, a_n$ are all real and a_0 is nonzero. We had to introduce complex numbers to provide a system within which all quadratic equations have solutions. Having done that, it is pleasing to know that polynomial equations of *any* degree have solutions within the system of complex numbers. Its solutions satisfy the property that if z_1 is a solution, so is its complex conjugate. I will not pursue this further here.

The simplest equation of the form (1.69) for a general value of n is $z^n = 1$. We will see that this equation has n solutions, which are all complex numbers. Obviously, $z^2 = 1$ has the two solutions $z = \pm 1$, and it is easily verified that $z^4 = 1$ has the solutions $z = \pm 1$ and $z = \pm i$. The solutions of $z^n = 1$ for some other small values of n can be found by elementary methods. (See, for example, Problem 1.5.4.) However, we will now see how we can find the solutions of $z^n = 1$ for *any* positive integer n. First we need a formal definition of angle.

Definition 1.5.4 The *angle BAC* in Figure 1.17, denoted by θ, is defined as the ratio of the length of the circular arc BC to the length of the line segment AB, the radius of the circular arc.   ■

The symbol θ denotes the Greek letter *theta*. Angles are sometimes measured in degrees, where 90 degrees corresponds to a right angle. This is not a fundamental way of measuring angles, since the number 90 is an arbitrary choice. Definition 1.5.4 gives the natural way of measuring angles, which

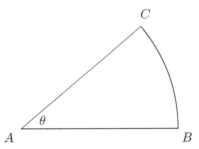

FIGURE 1.17. The angle BAC is defined as the ratio of the length of the circular arc BC to the length of the radius AB.

is called *radian* measure. In the case in which the circular arc BC has the same length as AB, the angle θ is one radian. The constant π is defined as the ratio of the circumference of a circle to its diameter, the symbol π being the Greek letter *pi*. It then follows from the definition of angle that the right angle has measure $\frac{\pi}{2}$ radians. The area bounded by AB, AC, and the circular arc BC is called a *sector*. Let $|AB| = |AC| = r$. Then the area of the sector ABC, which is obviously proportional to the angle θ, is equal to $\frac{1}{2}\theta r^2$. Thus, when $\theta = 2\pi$, we correctly obtain πr^2 for the area of the circle of radius r.

We will now define the sine and cosine functions, $\sin\theta$ and $\cos\theta$, with the aid of Figure 1.18, where P is a point on the circumference of the circle that has radius r, and whose center is the point O, with coordinates $(0,0)$. The point P has coordinates (x, y), and the line OP makes an angle θ with the x-axis. For the point P displayed in Figure 1.18, the x-coordinate is negative and the y-coordinate is positive. (More on this topic is given in Section 6.4.)

Note that *positive* angles are measured in a counterclockwise direction from the x-axis, and *negative* angles are measured in a clockwise direction. Then, for any real value of θ, we define

$$\sin\theta = \frac{y}{r} \quad \text{and} \quad \cos\theta = \frac{x}{r}. \tag{1.70}$$

The functions $\sin\theta$ and $\cos\theta$ are called *circular* functions, after the way they are defined in (1.70). It follows from (1.70) that the values of $\sin\theta$ and $\cos\theta$ are unchanged if we replace θ by $\theta + 2k\pi$, where k is any integer. We say that the sine and cosine functions are *periodic*, with period 2π. It also follows from (1.70), on putting $x = 0$ and $x = \frac{\pi}{2}$, that

$$\sin 0 = 0, \quad \cos 0 = 1, \quad \sin\frac{\pi}{2} = 1, \quad \cos\frac{\pi}{2} = 0. \tag{1.71}$$

We also see from (1.70) that for all real values of θ,

$$\sin(-\theta) = -\sin\theta \quad \text{and} \quad \cos(-\theta) = \cos\theta. \tag{1.72}$$

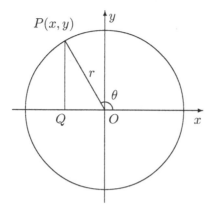

FIGURE 1.18. Circular functions: $\sin\theta = \frac{y}{r}$, $\cos\theta = \frac{x}{r}$.

In Figure 1.18 we note from Pythagoras's theorem that $x^2 + y^2 = r^2$. On dividing this last equation throughout by r^2, we see from (1.70) that

$$\cos^2\theta + \sin^2\theta = 1. \tag{1.73}$$

Following the usual custom, we have written $\cos^2\theta$ and $\sin^2\theta$ to denote $(\cos\theta)^2$ and $(\sin\theta)^2$, respectively.

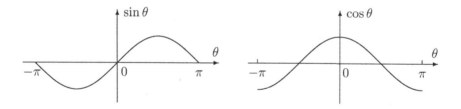

FIGURE 1.19. The functions $\sin\theta$ and $\cos\theta$ on the interval $[-\pi, \pi]$.

The graphs of the sine and cosine functions on the interval $[-\pi, \pi]$ are shown in Figure 1.19, and these shapes are copied endlessly to the left and to the right, following the periodic property of these functions. The graphs of the sine and cosine on $(-\infty, \infty)$ look the same. Indeed, we can see by considering how we defined the sine and cosine, using Figure 1.18, that

$$\sin\left(\theta + \frac{\pi}{2}\right) = \cos\theta. \tag{1.74}$$

We say that each graph is a *translation* of the other. The sine and cosine functions satisfy many identities, including the following two that we will use in our further study of complex numbers:

$$\sin(\alpha + \beta) = \sin \alpha \cos \beta + \cos \alpha \sin \beta, \tag{1.75}$$

$$\cos(\alpha + \beta) = \cos \alpha \cos \beta - \sin \alpha \sin \beta. \tag{1.76}$$

The identities (1.75) and (1.76), which hold for all real values of α and β, can be verified by elementary geometrical arguments.

Given any nonzero complex number $z = x + yi$, we can write

$$z = r \left(\frac{x}{r} + \frac{y}{r} i \right), \quad \text{where} \quad r = \sqrt{x^2 + y^2} = |z|.$$

Thus any nonzero complex number $z = x + yi$ can be expressed in the form

$$z = x + yi = r \left(\cos \theta + i \sin \theta \right), \tag{1.77}$$

where

$$\cos \theta = \frac{x}{r}, \quad \sin \theta = \frac{y}{r}, \quad \text{with} \quad r = \sqrt{x^2 + y^2}. \tag{1.78}$$

It is the standard practice to write the symbol i before $\sin \theta$, as on the right side of (1.77), instead of after it, as I have done in writing the general complex number as $x + yi$. We call $r \left(\cos \theta + i \sin \theta \right)$ the *polar* form of a complex number. Because of the periodicity of the sine and cosine functions, the choice of θ in (1.77) is not unique. We can obtain a unique value for θ by requiring that it satisfy the inequalities $-\pi < \theta \leq \pi$, and call this value of θ the *argument* of z. The following theorem is named after Abraham De Moivre (1667–1754).

Theorem 1.5.1 For any positive integer n we have

$$(\cos \theta + i \sin \theta)^n = \cos n\theta + i \sin n\theta \tag{1.79}$$

for all real values of θ.

Proof. We will use mathematical induction. The result is obviously true for $n = 1$. Let us assume that (1.79) holds for some value of $n \geq 1$. Then

$$(\cos \theta + i \sin \theta)^{n+1} = (\cos \theta + i \sin \theta)^n \cdot (\cos \theta + i \sin \theta), \tag{1.80}$$

and using (1.79), we obtain

$$(\cos \theta + i \sin \theta)^{n+1} = (\cos n\theta + i \sin n\theta) (\cos \theta + i \sin \theta) = X + Y i,$$

say, where

$$X = \cos n\theta \cos \theta - \sin n\theta \sin \theta \quad \text{and} \quad Y = \sin n\theta \cos \theta + \cos n\theta \sin \theta.$$

On putting $\alpha = n\theta$ and $\beta = \theta$ in (1.75) and (1.76), we find that X and Y can be written more simply as $X = \cos(n + 1)\theta$ and $Y = \sin(n + 1)\theta$.

Thus (1.79) holds when n is replaced by $n+1$, and this completes the proof by mathematical induction.   ■

Now let us find the solutions of the equation $z^n = 1$. It follows from (1.64) that $|z^n| = |z|^n$, so that $|z| = 1$. Thus $z = \cos\theta + i\sin\theta$, for some value of θ. Then, by Theorem 1.5.1, we obtain

$$z^n = \cos n\theta + i\sin n\theta = 1, \tag{1.81}$$

so that

$$\cos n\theta = 1 \quad \text{and} \quad \sin n\theta = 0. \tag{1.82}$$

In view of (1.71) and the periodicity of the sine and cosine functions, we deduce from (1.82) that $n\theta = 2k\pi$, where k is any integer. Consequently, $\theta = 2k\pi/n$, and all solutions of (1.81) are of the form

$$z = \cos\frac{2k\pi}{n} + i\sin\frac{2k\pi}{n}, \tag{1.83}$$

where k is an integer. The choice of $k = 0, 1, \ldots, n-1$ gives n distinct solutions, and because of the periodicity of the sine and cosine, no further choice of k yields any more solutions. We call the n solutions of $z^n = 1$ the nth roots of unity. They are equally spaced on the circumference of the circle $|z| = 1$, one root always being on the x-axis, corresponding to the root $z = 1$. The case $n = 7$ is illustrated in Figure 1.20.

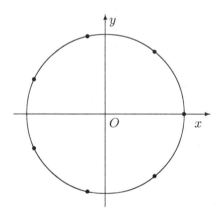

FIGURE 1.20. The seventh roots of unity.

Problem 1.5.1 Verify that the equation $z^2 - z + 1 = 0$ has solutions

$$z_1 = \frac{1}{2} + \frac{\sqrt{3}}{2}i \quad \text{and} \quad z_2 = \frac{1}{2} - \frac{\sqrt{3}}{2}i.$$

Evaluate $|z_1|$, $|z_2|$, $z_1 z_2$, and $|z_1 z_2|$.

Problem 1.5.2 With z_1 and z_2 as defined in Problem 1.5.1, show that

$$z_1^3 = z_2^3 = -1.$$

Problem 1.5.3 Show that the equation $z^2 + z + 1 = 0$ has solutions

$$z_3 = -\frac{1}{2} + \frac{\sqrt{3}}{2}i \quad \text{and} \quad z_4 = -\frac{1}{2} - \frac{\sqrt{3}}{2}i.$$

Verify that $z_1^2 = z_3$ and $z_2^2 = z_4$, where z_1 and z_2 are defined in Problem 1.5.1, and that $z_3^3 = z_4^3 = 1$.

Problem 1.5.4 Show that

$$z^6 - 1 = (z^3 - 1)(z^3 + 1) = (z - 1)(z^2 + z + 1)(z + 1)(z^2 - z + 1).$$

Using the results obtained in the three problems above, show that the set of solutions of $z^6 - 1 = 0$ is $\{z_1, z_1^2, z_1^3, z_1^4, z_1^5, z_1^6\}$, where $z_1 = \frac{1}{2} + \frac{\sqrt{3}}{2}i$.

Problem 1.5.5 Deduce from the inequality $(x_1 y_2 - x_2 y_1)^2 \geq 0$ that

$$(x_1 x_2 + y_1 y_2)^2 \leq \left(x_1^2 + y_1^2\right)\left(x_2^2 + y_2^2\right)$$

and thus verify that

$$|x_1 x_2 + y_1 y_2| \leq |z_1| |z_2|,$$

so that

$$2x_1 x_2 + 2y_1 y_2 \leq 2|z_1| |z_2|,$$

where $z_1 = x_1 + y_1 i$ and $z_2 = x_2 + y_2 i$. By adding $x_1^2 + y_1^2 + x_2^2 + y_2^2$ to both sides of the latter inequality, show that

$$|z_1 + z_2|^2 \leq (|z_1| + |z_2|)^2.$$

Problem 1.5.6 Verify that

$$|z_1 - z_2|^2 = (z_1 - z_2)(\bar{z}_1 - \bar{z}_2) = |z_1|^2 - 2\operatorname{Re}(z_1 z_2) + |z_2|^2$$

and hence prove that

$$|z_1 - z_2|^2 \geq (|z_1| - |z_2|)^2.$$

Problem 1.5.7 Let $p(z)$ denote a polynomial of degree n in z with real coefficients. Show that the complex conjugate of $p(z)$ is $p(\bar{z})$. Deduce that $p(\alpha) = 0$ if and only if $p(\bar{\alpha}) = 0$, and that the complex zeros of $p(z)$ occur in conjugate pairs.

2
Numbers, Numbers Everywhere

Wherever there is number, there is beauty.

Proclus (410–485)

2.1 Different Bases

Our familiar number system can be described as a *positional* number system with *base* 10, or more simply but less informatively as the decimal system. Every positive integer n can be expressed uniquely in the form

$$n = a_k \times 10^k + a_{k-1} \times 10^{k-1} + \cdots + a_2 \times 10^2 + a_1 \times 10 + a_0. \qquad (2.1)$$

We denote n by the string of digits $a_k a_{k-1} \ldots a_1 a_0$, which we call its decimal form. For example,

$$56852 = 5 \times 10^4 + 6 \times 10^3 + 8 \times 10^2 + 5 \times 10 + 2.$$

Recall that 10^2 means 10×10, 10^3 denotes the product $10 \times 10 \times 10$, and so on. The quantities 2 and 3 that appear in 10^2 and 10^3 are called *exponents*. Each number $a_0, a_1, \ldots, a_{k-1}$ that occurs in (2.1) is an integer between 0 and 9, and a_k is an integer between 1 and 9, that is,

$$0 \le a_j \le 9 \quad \text{for} \quad 0 \le j < k, \quad \text{and} \quad 1 \le a_k \le 9.$$

We call a_k the most-significant digit of n. It gives us a rough idea of the size of n, for we have

$$a_k \times 10^k \le n < (a_k + 1) \times 10^k.$$

For example,

$$5 \times 10^4 \le 56852 < 6 \times 10^4.$$

Likewise, we call a_0 the least-significant digit of n.

It has been said that we count in tens because we have ten *digits* on our two hands. Indeed, the English word digit is derived from the Latin word *digitus*, meaning finger. However, any other positive integer greater that 1 can be used instead of ten as the base of a number system.

Example 2.1.1 Let us express the decimal number 149 in base 3. First we divide 149 by 3 to give the quotient 49 and remainder 2. We write this in the form

$$149 = 49 \times 3 + 2, \tag{2.2}$$

showing that 2 is the least-significant digit of the number 149 in base 3. Then we divide 49 by 3 to give the quotient 16 and remainder 1, which we express as

$$49 = 16 \times 3 + 1. \tag{2.3}$$

We continue this process of repeatedly dividing by 3 and finding the remainder, obtaining

$$16 = 5 \times 3 + 1, \quad \text{and} \quad 5 = 3 + 2.$$

On combining the last two equations, we find that

$$16 = (3 + 2) \times 3 + 1 = 3^2 + 2 \times 3 + 1,$$

and hence, using (2.3),

$$49 = (3^2 + 2 \times 3 + 1) \times 3 + 1 = 3^3 + 2 \times 3^2 + 3 + 1.$$

Finally, we have from (2.2) that

$$149 = (3^3 + 2 \times 3^2 + 3 + 1) \times 3 + 2 = 3^4 + 2 \times 3^3 + 3^2 + 3 + 2.$$

Since the coefficients of 3^4, 3^3, 3^2, 3, and 1 are 1, 2, 1, 1, and 2, respectively, we write $(12112)_3$ to represent the decimal number 149, and we have written the suffix 3 after the number 12112 to emphasize that it is expressed in base 3. Similarly we could write 149 as $(149)_{10}$, and we have shown that $(149)_{10} = (12112)_3$. ■

It is important to realize that a number exists independently of the way we record it on the page. We can express a given positive integer in any base $b \ge 2$. Let us construct a form $(a_k a_{k-1} \ldots a_1 a_0)_b$ for the number n_0 in base b, where $b \ge 2$. We use the *division algorithm*, which I now define.

Definition 2.1.1 (The division algorithm) Given any two integers a and b, with $b > 0$, we can always find unique integers q and r such that

$$a = qb + r, \quad \text{where} \quad 0 \le r < b. \quad \blacksquare \tag{2.4}$$

Remark 2.1.1 My dictionary defines *algorithm* as a process or set of rules used for calculation or problem-solving, especially with a computer. $\blacksquare$

See Grimaldi [10] or Phillips [14] for a justification of the division algorithm. Given the base $b > 1$ and a positive integer n_0, we apply the division algorithm to give a unique quotient n_1 and a unique remainder a_0 that satisfy

$$n_0 = n_1 b + a_0, \quad \text{where} \quad n_1 \ge 0 \quad \text{and} \quad 0 \le a_0 < b. \tag{2.5}$$

If $n_1 > b$, we apply the division algorithm to n_1, giving

$$n_1 = n_2 b + a_1, \quad \text{where} \quad 0 \le a_1 < b. \tag{2.6}$$

Following (2.6), if $n_2 > b$, we determine n_3 and a_2 such that

$$n_2 = n_3 b + a_2, \quad \text{where} \quad 0 \le a_2 < b.$$

We continue to apply the division algorithm, finding quotients n_j and remainders a_{j-1} until we reach a stage at which, having computed

$$n_{k-1} = n_k b + a_{k-1}, \quad \text{where} \quad 0 \le a_{k-1} < b, \tag{2.7}$$

we find that n_k is less than b. (Since the sequence of numbers n_0, n_1, and so on, is decreasing, there must be a positive integer k such that $n_k < b$.) We then terminate the process and define $a_k = n_k$. Since $n_{k-2} = n_{k-1}b + a_{k-2}$, we can replace n_{k-1} by the expression given for it in (2.7) and so obtain

$$n_{k-2} = a_k b^2 + a_{k-1} b + a_{k-2}. \tag{2.8}$$

Similarly, we write $n_{k-3} = n_{k-2}b + a_{k-3}$, substitute for n_{k-2} its expression given in (2.8), and obtain

$$n_{k-3} = a_k b^3 + a_{k-1} b^2 + a_{k-2} b + a_{k-3}.$$

We continue until we finally obtain

$$n_0 = a_k b^k + a_{k-1} b^{k-1} + \cdots + a_2 b^2 + a_1 b + a_0, \tag{2.9}$$

and we will denote this by $a_k a_{k-1} \ldots a_1 a_0$, with the most-significant digit, a_k, written down first. Following what we did above for the bases 10 and 3, we will write this in the form $(a_k a_{k-1} \ldots a_1 a_0)_b$ when we need to emphasize that the *base* is b. The expression on the right of (2.9) is a polynomial in b of degree k.

+	1	2	3	4
1	2	3	4	10
2	3	4	10	11
3	4	10	11	12
4	10	11	12	13

TABLE 2.1. Addition table for base-five arithmetic.

Given two positive integers m and n that are both expressed in base b, we can express their sum in base b. This entails adding corresponding digits, beginning with the least-significant, and adding any "carry" digit, as we do in base-10 arithmetic. When we are working in base-b arithmetic for the first time, even this simple task is unfamiliar to us, and it is helpful to construct a table for adding single digits. An addition table for base-five arithmetic is given in Table 2.1.

Example 2.1.2 Working in base-five arithmetic, let us find the sum of $(342)_5$ and $(421)_5$. We have

$$\begin{array}{r} 421 \\ +342 \\ \hline 1313 \end{array}$$

and the result is $(1313)_5$. Note that there is no carry when we add the first digits, 2 and 1. Next we add 4 and 2 (giving $1 + 5$) to give 1 plus a carry digit. Finally, we add 3 and 4 plus the carry digit (giving $3 + 5$) to give 3 plus a carry digit, and the carry digit is written down in the next place to give the most-significant digit in the sum.

As a check on our calculation, we can convert $(342)_5$ and $(421)_5$ into their decimal equivalents, which are 97 and 111, respectively. The sum of these two decimal numbers is 208, and we can verify that this is indeed the decimal equivalent of $(1313)_5$. ■

×	1	2	3	4
1	1	2	3	4
2	2	4	11	13
3	3	11	14	22
4	4	13	22	31

TABLE 2.2. Multiplication table for base five-arithmetic.

Let us consider the *multiplication* of two numbers written in any base b. We follow the same method that we use in base-10 arithmetic. We need to construct a table showing the result of multiplying any two single digits together in the chosen base. For example, Table 2.2 gives such a multiplication table for base-five arithmetic. I have omitted multiplications involving zero from the table.

Example 2.1.3 Let us multiply $(32)_5$ by $(23)_5$. The scheme of calculations is set out as follows:

$$
\begin{array}{r}
32 \\
\times\ 23 \\
\hline
201 \\
114 \\
\hline
1341
\end{array}
$$

The number $(201)_5$ is the result of multiplying $(32)_5$ by 3, which is the least-significant digit of $(23)_5$, and $(114)_5$ is the result of multiplying $(32)_5$ by 2. Note how we have shifted 114 one place to the left. This makes it equivalent to the number $(1140)_5$. As a check on our calculation, we note that $(32)_5$ and $(23)_5$ have the decimal equivalents 17 and 13, respectively, and we can verify that the decimal number $17 \times 13 = 221$ does indeed correspond to $(1341)_5$. ■

$\times$	1	2	3	4	5	6	7	8	9	δ	ϵ
1	1	2	3	4	5	6	7	8	9	δ	ϵ
2	2	4	6	8	δ	10	12	14	16	18	1δ
3	3	6	9	10	13	16	19	20	23	26	29
4	4	8	10	14	18	20	24	28	30	34	38
5	5	δ	13	18	21	26	2ϵ	34	39	42	47
6	6	10	16	20	26	30	36	40	46	50	56
7	7	12	19	24	2ϵ	36	41	48	53	5δ	65
8	8	14	20	28	34	40	48	54	60	68	74
9	9	16	23	30	39	46	53	60	69	76	83
δ	δ	18	26	34	42	50	5δ	68	76	84	92
ϵ	ϵ	1δ	29	38	47	56	65	74	83	92	$\delta 1$

TABLE 2.3. Multiplication table for base-twelve arithmetic.

If the base b is greater than ten, we need symbols to denote the additional digits required. In Table 2.3, the multiplication table for base twelve, the symbols δ and ϵ (the Greek letters *delta* and *epsilon*) are used to represent ten and eleven, respectively.

The base with the simplest arithmetic is, of course, base two. This gives the *binary* system, which requires only two digits, 0 and 1. For addition we need only know that $0 + 0 = 0$, $0 + 1 = 1$, and $1 + 1 = 10$, and for multiplication, that $0 \times 0 = 0$, $0 \times 1 = 0$, and $1 \times 1 = 1$.

Example 2.1.4 Let us convert the decimal numbers 23 and 29 into binary numbers, and find their sum and product. We have $23 = (10111)_2$

and $29 = (11101)_2$. We find that their sum is $(110100)_2$, and their product is $(1010011011)_2$. It is easy to get lost when we are adding up the contributions to the product! We can check that the binary numbers for the sum and product have the decimal equivalents 52 and 667, and that these are correct. ■

I have an old-fashioned balance for weighing things. On one side of the balance is the pan into which one puts the material to be weighed, such as flour, and on the other side is the platform where one puts the weights. There are twelve weights, of

$$1, \ 2, \ 2, \ 5, \ 10, \ 20, \ 20, \ 50, \ 100, \ 200, \ 200, \ 500$$

grams. It is easy to check that we can weigh any whole number of grams from 1 to 10 with the four weights of 1, 2, 2, and 5 grams, and the weights of 10, 20, 20, and 50 grams can weigh any multiple of 10 grams from 10 to 100. In this way we can see that using some or all of the set of twelve weights, we can weigh any whole number of grams from 1 to 1110. However, it follows from the representation of numbers in the binary system that any integer m between 1 and $2^n - 1$ can be expressed uniquely in binary form with at most n digits; that is, m can be expressed as a sum of some or all of the numbers $1, 2, 2^2, \ldots, 2^{n-1}$. Note also that

$$2^n - 1 = (111\ldots1)_2, \tag{2.10}$$

where there are n digits 1 on the right of (2.10). Thus, if I had a set of ten weights of

$$1, \ 2, \ 2^2, \ 2^3, \ 2^4, \ 2^5, \ 2^6, \ 2^7, \ 2^8, \ 2^9$$

grams, I could weigh any whole number of grams up to $2^{10} - 1 = 1023$.

A number in base three contains only the digits 0, 1, and 2. In each instance where the digit 2 occurs we could write

$$2 \times 3^j = (3-1) \times 3^j = 3^{j+1} - 3^j.$$

Hence any positive integer can be expressed as the sum of certain powers of three *minus* the sum of certain other powers of three. For example,

$$100 = 3^4 + 3^3 - 3^2 + 3^0.$$

Suppose I had a balance in which weights may be placed on either side. Thus any weight placed on the same side as the material to be weighed acts like a negative weight. Then with only seven weights, of

$$1, \ 3, \ 3^2, \ 3^3, \ 3^4, \ 3^5, \ 3^6$$

grams, I could weigh any whole number of grams up to

$$1 + 3 + 3^2 + \cdots + 3^6 = 1093.$$

There is a game called nim whose analysis depends on an understanding of base-two arithmetic. It is played, using a number of counters arranged in heaps, by two players whom we will call M and N. You can think of them as the mathematician and the nonmathematician. The players make alternate moves, and a move involves removing counters from one heap only. Any number of counters can be removed, from at least one up to the whole heap, but only from one heap. The player who removes the last counter wins the game. The case where there is just one counter in each heap is easily analyzed: if there is an odd number of heaps, the winner will be the one who plays first, and if there is an even number of heaps, the winner will be the one who plays second. Another case that is easily analyzed is that of just two heaps. Suppose that when it is M's turn, the two heaps have a different number of counters. M should remove counters from the larger heap to leave the same number in each heap. Then, whatever N does, the two heaps will again have a different number of counters for M's next turn. M then repeats this strategy of leaving two equal heaps, and ultimately must win.

Let us write $(n_1, n_2, \cdots, n_k)$ to denote a configuration in which we have k heaps, one containing n_1 counters, another containing n_2 counters, and so on. We can let $n_1 \leq n_2 \leq \cdots \leq n_k$, since the *order* of the heaps does not matter. Let me explain M's strategy in a particular case. Suppose that M is faced with the configuration $(3, 6, 9, 14, 21, 29)$. M writes down these six numbers in binary form and adds up the numbers in each binary place, as shown in Table 2.4.

$$
\begin{array}{cc}
00011 & 3 \\
00110 & 6 \\
01001 & 9 \\
01110 & 14 \\
10101 & 21 \\
11101 & 29 \\
\hline
23434 \\
\circ \bullet \circ \bullet \circ
\end{array}
$$

TABLE 2.4. An odd configuration in the game of nim.

Note that extra zeros have been inserted, where needed, to give all the binary numbers the same number of digits, and the five columns of 0's and 1's have been summed separately. We will call such a configuration *even* if *every* column sum is even, and otherwise, we will call the configuration *odd*. Thus the above configuration is odd, since two of the column sums are odd (both 3). For the sake of clarity, let me emphasize that a configuration is odd if one or more column sum is odd. M's strategy is to transform an odd configuration into an even one. M selects a row in which there is a 1 in the most-significant column of those whose sum is odd. In this case it is the second column from the left, corresponding to the 2^3 digit. In general, there must always be such a row in an odd configuration, since a column whose sum is odd must contain at least one 1. Here, M could select the third row, 01001, and replace it by 00011, which corresponds to replacing 9 by 3. Thus M's move is to take six counters from the heap containing nine. This has the effect of changing only the two odd sums, and replacing them with even sums, and so changes the configuration from odd to even. M could equally have selected the fourth or the sixth rows to be changed, since these also have a 1 in the second column. The 1 is changed to a 0 and the parity of the digit in the fourth column is changed, meaning that a 0 is replaced by 1 and a 1 is replaced by 0. Note that this change always reduces the size of the number that is altered, and in general, an odd configuration can always be changed to an even one by removing an appropriate number of counters from one of the heaps.

Beginning with an even configuration, N has to change at least one digit, and so inevitably N's move turns an even configuration into an odd one, which M can make even again. If, at some stage, M's move leaves the even configuration consisting of all zeros, then all the counters have been removed and M has won. Otherwise, N has an even configuration, which must therefore mean that two or more heaps remain, and N cannot win at this move. Someone must win after a finite number of moves, since each move reduces the number of counters. M, being the mathematician, is the hero of this story, who can always turn an odd configuration into an even one. Consequently, M will lose only if the configuration happens to be odd on N's first turn and N plays as cleverly as M!

Problem 2.1.1 Write down addition and multiplication tables for base b, for all values of b between 2 and 10.

Problem 2.1.2 Add and multiply the numbers $(2063)_7$ and $(1542)_7$, working entirely in base-7 arithmetic. Check your results by converting the two numbers into their representations in base 10 and repeating the two calculations.

Problem 2.1.3 Consider the nim configuration $(3, 6, 9, 14, 21, 29)$, which is displayed in Table 2.4. In the text, we saw that this odd configuration could be changed to an even one by taking 6 counters from the heap of 9. Show that an even configuration may also be achieved by taking 10 counters from the heap of 14, or 6 counters from the heap of 29.

Problem 2.1.4 The last binary digit of any number of the form $2n$ is 0, and if we replace this 0 by 1, we transform the binary representation of $2n$ into that for $2n+1$. Hence show that the nim configuration for $(1, 2n, 2n+1)$ is even.

Problem 2.1.5 Show that nim configurations of the forms $(2,3,4,5)$ and $(n,7-n,7)$ for $1 \leq n \leq 3$ are even.

Problem 2.1.6 Let $(m_1,\ldots,m_r)$ and $(n_1,\ldots,n_s)$ be even configurations in the game of nim. Show that $(m_1,\ldots,m_r,n_1,\ldots,n_s)$ is also an even configuration.

2.2 Congruences

Sometimes, in working with numbers, we count up to some positive integer n, and then begin again from zero. For example, with the days of the week, we "count" Sunday, Monday, Tuesday, Wednesday, Thursday, Friday, Saturday, and then begin again with Sunday. Similarly, in a clock with numbers 1 to 12, a cycle of 12 hours is repeated, and there is a cycle of 24 hours in the 24-hour clock.

Let me give one simple example, for which we don't need any special notation, concerning days of the week. Some time ago I observed from reading the local newspaper that golden wedding anniversaries (fifty years) often happened on a Saturday. Now, if a couple are married on a Saturday, the first anniversary of their wedding will fall on a Sunday, unless a leap year day (29 February) occurs in the first year of their marriage. Of course, this is because 365 has a remainder 1 when divided by 7, and adding one day to Saturday gives Sunday. On each wedding anniversary we have to add one day, except that each time 29 February comes around, we have to add two days, because 366 has a remainder 2 when divided by 7. So after 50 years we need to add $L + 50$ days, where L is the number of times 29 February has occurred. Since 50 has a remainder 1 when divided by 7, adding $L + 50$ is the same as adding $L + 1$. (In what follows, let us assume that the period concerned does not include years such as 1900 and 2100, which are *not* leap years.) Then, over the 50-year period, there must be either 12 or 13 leap year days. Thus $L + 1$ has the value 13 or 14, and so the 50th wedding anniversary must be either on the same day of the week as the wedding, or on the day before that. Let me add that I have a personal interest in this problem because I was born on a Saturday and my fiftieth birthday was also on a Saturday.

We use the notation $n \mid m$ to denote n divides m. In other words, m is an integer multiple of n, where n and m are integers. We now define the notion of *congruence*.

Definition 2.2.1 We write

$$x \equiv u \,(\mathrm{mod}\, n) \tag{2.11}$$

to mean that the integers x and u have the same remainder when each is divided by the positive integer n. ■

Theorem 2.2.1 If $x \equiv u \pmod{n}$, then $n \mid x - u$.

Proof. See Problem 2.2.1.   ■

We read (2.11) as "x is congruent to u modulo n," and refer to n as the *modulus*. For example,

$$366 \equiv 2 \pmod{7} \quad \text{and} \quad 50 \equiv 1 \pmod{7}.$$

The concept of a *congruence relation* was introduced and developed by C. F. Gauss. Congruences satisfy the three basic properties

$$x \equiv x \pmod{n}, \tag{2.12}$$
$$x \equiv y \pmod{n} \;\Rightarrow\; y \equiv x \pmod{n}, \tag{2.13}$$
$$x \equiv y \pmod{n} \;\text{and}\; y \equiv z \pmod{n} \;\Rightarrow\; x \equiv z \pmod{n}, \tag{2.14}$$

where the symbol $\Rightarrow$ means *implies that*. Starting with the definition of congruence, it is not hard to show that these properties hold. For example, let us verify (2.14). If

$$x \equiv y \pmod{n} \quad \text{and} \quad y \equiv z \pmod{n},$$

then it follows from the definition of congruence that there exist integers r and s such that

$$x - y = rn \quad \text{and} \quad y - z = sn,$$

and hence we have

$$x - z = (x - y) + (y - z) = (r + s)n.$$

Since this last equation shows that $n \mid x - z$, it follows that $x \equiv z \pmod{n}$, thus justifying (2.14).

Remark 2.2.1 The three properties (2.12), (2.13), and (2.14) are called the reflexive, symmetric, and transitive properties, respectively. More generally in mathematics, any relation between pairs of members of a given set that satisfies these three properties is called an *equivalence relation*. It may be helpful to state a relation that is *not* an equivalence relation. One such relation is $x \leq y$, meaning "x is less than or equal to y." For although it is true that $x \leq x$ and that

$$x \leq y \quad \text{and} \quad y \leq z \;\Rightarrow\; x \leq z,$$

it is not true that

$$x \leq y \;\Rightarrow\; y \leq x.$$

Thus the "less than or equal to" relation is *not* an equivalence relation, since the symmetric property does not hold. However, the relation defined by "A's birthday falls on the same day of the week as B's birthday" *is* an equivalence relation.   ■

Suppose now that

$$x \equiv u \,(\mathrm{mod}\ n) \qquad \text{and} \qquad y \equiv v \,(\mathrm{mod}\ n). \tag{2.15}$$

Then we see from Theorem 2.2.1 that there are integers r and s such that

$$x - u = rn \quad \text{and} \quad y - v = sn.$$

It follows that

$$(x + y) - (u + v) = (x - u) + (y - v) = (r + s)n,$$

and hence

$$x + y \equiv u + v \,(\mathrm{mod}\ n). \tag{2.16}$$

We also have

$$(xy - uv) = (x - u)y + u(y - v) = (ry + us)n,$$

and since $ry + us$ is an integer, we obtain

$$xy \equiv uv \,(\mathrm{mod}\ n). \tag{2.17}$$

Thus (2.16) and (2.17) both follow from (2.15).

Example 2.2.1 By choosing $y = x$ and $v = u$ in (2.17), we see that

$$x \equiv u \,(\mathrm{mod}\ n) \quad \Rightarrow \quad x^2 \equiv u^2 \,(\mathrm{mod}\ n),$$

where n is any positive integer. If we now put $y = x^2$ and $v = u^2$ in (2.17), we have

$$x \equiv u \,(\mathrm{mod}\ n) \quad \Rightarrow \quad x^3 \equiv u^3 \,(\mathrm{mod}\ n).$$

More generally, using mathematical induction, we obtain

$$x \equiv u \,(\mathrm{mod}\ n) \quad \Rightarrow \quad x^k \equiv u^k \,(\mathrm{mod}\ n), \tag{2.18}$$

for any positive integer k. ∎

Let us accept without proof, at this stage, that if a prime p divides cd, where c and d are positive integers, then p divides at least one of the integers c and d. (See Theorem 2.4.3.) This immediately yields an important result involving congruences, that if p is a prime that does not divide a, then

$$ab \equiv ac \,(\mathrm{mod}\ p) \quad \Rightarrow \quad b \equiv c \,(\mathrm{mod}\ p). \tag{2.19}$$

For

$$ab \equiv ac \,(\mathrm{mod}\ p) \quad \Rightarrow \quad p \mid ab - ac \quad \Rightarrow \quad p \mid a(b - c).$$

Since the prime p does not divide a, it must divide $b - c$, and so we have $b \equiv c \,(\mathrm{mod}\ p)$, which justifies (2.19).

+	1	2	3	4	5	6
1	2	3	4	5	6	0
2	3	4	5	6	0	1
3	4	5	6	0	1	2
4	5	6	0	1	2	3
5	6	0	1	2	3	4
6	0	1	2	3	4	5

×	1	2	3	4	5	6
1	1	2	3	4	5	6
2	2	4	6	1	3	5
3	3	6	2	5	1	4
4	4	1	5	2	6	3
5	5	3	1	6	4	2
6	6	5	4	3	2	1

TABLE 2.5. Addition and multiplication tables modulo 7.

All of the above properties of congruences still look correct if we replace the symbol $\equiv$ by $=$, and we can thank Gauss for choosing such an appropriate symbol for congruence. The similarity to ordinary equality makes it easy for us to carry out calculations using congruences.

If $x \equiv a \,(\mathrm{mod}\; n)$, we say that a is a *residue* of x modulo n. The set of all residues modulo n of a given number is called a *residue class*. There are n residue classes modulo n, namely those that are congruent to $0, 1, \ldots, n-1$. Given any positive integer n, we could write down addition and multiplication tables for all pairs of residues modulo n. Table 2.5 contains addition and multiplication tables modulo 7. Multiplication tables of residues are more interesting than addition tables. It follows from (2.19) that in a multiplication table of residues modulo p, where p is any prime, the numbers in any row are all different. This means that the numbers in any row must be just the numbers $1, 2, \ldots, p - 1$, in some order, as we see in Table 2.5 for the case $p = 7$. Thus it follows that for any residue a between 1 and 6,

$$(a)(2a)(3a)(4a)(5a)(6a) \equiv 1 \cdot 2 \cdot 3 \cdot 4 \cdot 5 \cdot 6 \,(\mathrm{mod}\; 7), \tag{2.20}$$

and since 7 does not divide the number on the right of (2.20), we can deduce from (2.19) that

$$a^6 \equiv 1 \,(\mathrm{mod}\; 7). \tag{2.21}$$

We can extend this result from the prime 7 to any prime. (Note that the case for the prime $p = 2$ is very simple, and therefore we need only to consider odd primes.) For any odd prime p we have, as we have in (2.20) for the special case $p = 7$,

$$(a)(2a) \cdots (p - 1)a \equiv 1 \cdot 2 \cdots (p - 1) \,(\mathrm{mod}\; p), \tag{2.22}$$

for any integer a such that $1 \leq a \leq p-1$, and so we can deduce from (2.19) that

$$a^{p-1} \equiv 1 \,(\mathrm{mod}\; p). \tag{2.23}$$

If we use (2.18), we can see that this result is valid for all values of a that are not congruent to zero modulo p. Thus we have proved the following theorem, which is due to Pierre de Fermat.

Theorem 2.2.2 (Fermat's Little Theorem) For any prime number p and and any positive integer a that is not divisible by p,

$$a^{p-1} \equiv 1 \,(\mathrm{mod}\ p). \qquad \blacksquare \qquad\qquad (2.24)$$

This famous theorem is usually called Fermat's little theorem. This may seem rather disparaging, but Theorem 2.2.2 is certainly little in comparison with Fermat's last theorem, to which I referred in Section 1.3.

Problem 2.2.1 Let $x \equiv u \,(\mathrm{mod}\ n)$. Write $x = an + c$ and $u = bn + c$, where a, b, and c are integers, with $0 \le c < n$. Deduce that $n \mid x - u$.

Problem 2.2.2 Verify directly from Definition 2.2.1 that if $a \equiv 1 \,(\mathrm{mod}\ 4)$ and $b \equiv 1 \,(\mathrm{mod}\ 4)$, then $ab \equiv 1 \,(\mathrm{mod}\ 4)$. Deduce that the product of any number of factors of the form $4n + 1$ is also of this form.

Problem 2.2.3 Show that $x^2 \equiv 0 \,(\mathrm{mod}\ 4)$ if x is even and $x^2 \equiv 1 \,(\mathrm{mod}\ 4)$ if x is odd.

Problem 2.2.4 Deduce from Fermat's little theorem that

$$10^6 \equiv 1 \,(\mathrm{mod}\ 7)$$

and hence show that

$$10^{100} \equiv 10^4 \,(\mathrm{mod}\ 7).$$

Deduce that 10^{100} has a remainder 4 when divided by 7. The very large number 10^{100} is called a *googol*.

Problem 2.2.5 What is the remainder when a googol is divided by 13?

Problem 2.2.6 Consider the equations for primitive Pythagorean triples (see Section 1.3),

$$a = 2uv, \quad b = u^2 - v^2, \quad c = u^2 + v^2,$$

where u and v are positive integers having no common factor greater than 1, and $u+v$ is odd. Verify that a is divisible by 4. Check that in any column of Table 1.4, one of a, b, and c is divisible by 3 and one is divisible by 5. Prove that this holds for all triples by writing down the residue classes of a, b, and c corresponding to all possible residue classes of u and v modulo 3, and do the same for the residue classes modulo 5. Deduce that $abc \equiv 0 \,(\mathrm{mod}\ 60)$.

2.3   Continued Fractions

We begin this section by restating the division algorithm (see Definition 2.1.1), which we used in Section 2.1. Given any two integers a and b, with $b > 0$, we can always find unique integers q and r such that

$$a = qb + r, \quad \text{where} \quad 0 \le r < b. \tag{2.25}$$

If $b > a \ge 0$, we need to choose the values $q = 0$ and $r = a$. If $0 < b \le a$, we divide a by b to obtain a positive integer q as the quotient, leaving a remainder r that satisfies the inequalities shown in (2.25). The division algorithm has an obvious geometrical interpretation when a and b are both positive, for we can think of a and b as lengths, and thus we could allow a and b to be any positive numbers, not necessarily integers. In applying the division algorithm, we are using b as a means of measuring the length a, determining how many units of length b are contained in a, and how much is left over.

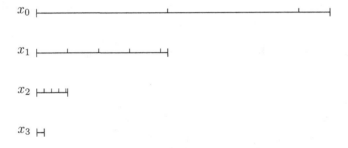

FIGURE 2.1. Repeated application of the division algorithm.

In what follows, we will apply the division algorithm repeatedly, and it will be helpful to modify our notation. Let us apply the division algorithm to the two real numbers x_0 and x_1, where $x_0 > x_1 > 0$, giving

$$x_0 = q_1 x_1 + x_2, \quad \text{where} \quad 0 \le x_2 < x_1 \tag{2.26}$$

and q_1 is a positive integer. Unless $x_2 = 0$, which will happen if and only if x_0 is an exact multiple of x_1, we can apply the division algorithm to the two real numbers x_1 and x_2, to give

$$x_1 = q_2 x_2 + x_3, \bullet \text{ where} \quad 0 \le x_3 < x_2 \tag{2.27}$$

and q_2 is a positive integer. If $x_3 > 0$, we can continue the process by applying the division algorithm to x_2 and x_3, and so on. Figure 2.1 illustrates this process for the case $q_1 = 2$, and $q_2 = q_3 = 4$. Obviously, in the general case, there are two possible outcomes. One possibility is that we obtain a zero remainder at some stage, and the sequence of equations that begins with (2.26) and (2.27) terminates after, say, k divisions with

$$x_{k-1} = q_k x_k, \tag{2.28}$$

where q_k is a positive integer. Since $x_k < x_{k-1}$, the final quotient q_k must be an integer greater than 1. The other possible outcome of the process that begins with (2.26) and (2.27) is that it never comes to an end.

Let us analyze what happens when the above process terminates at stage k with the equation (2.28). For $j < k$, the jth equation is

$$x_{j-1} = q_j x_j + x_{j+1},$$

and we can divide throughout by x_j to give

$$\frac{x_{j-1}}{x_j} = q_j + \frac{x_{j+1}}{x_j}. \tag{2.29}$$

Now let us assume that x_j/x_{j+1} is a positive rational number, so that

$$\frac{x_j}{x_{j+1}} = \frac{a}{b},$$

where a and b are positive integers with no common factor. Then we see from (2.29) that

$$\frac{x_{j-1}}{x_j} = q_j + \frac{b}{a} = \frac{q_j a + b}{a},$$

and it follows that x_{j-1}/x_j is a positive rational number. This is the key step in this argument: beginning with the *assumption* that x_j/x_{j+1} is a positive rational number, we have *deduced* that x_{j-1}/x_j is a positive rational number. However, we see from (2.28) that x_{k-1}/x_k is a positive rational number. (In fact, it is a positive integer, a special case of a positive rational number.) Thus, working back from the kth equation to the first equation, applying the same argument as we used above with the jth equation, we deduce in turn that all the ratios, from x_{k-1}/x_k to x_0/x_1, are positive rational numbers. This argument relies on mathematical induction.

We have just proved that if the process terminates, the ratio x_0/x_1 is a positive rational number. Conversely, suppose that x_0/x_1 is a positive rational number. Can we prove that the process must terminate? This is not difficult, for it will suffice to consider the case in which x_0 and x_1 are positive integers. (For we can multiply x_0 and x_1 by some suitable positive integer to give integers, and this is equivalent to multiplying all the equations by this positive integer.) Given that x_0 and x_1 are positive integers, and since it follows from the above equations that $x_0 > x_1 > x_2$, and so on, we see that this decreasing sequence of positive integers must terminate.

If x_0/x_1 is an *irrational number*, which means that x_0/x_1 is *not* a rational number, then the process cannot terminate, because we saw above that if the process terminates, x_0/x_1 must be rational. We now give an example in which the process terminates.

Example 2.3.1 With $x_0 = 97$ and $x_1 = 29$, we construct x_2, x_3, and x_4 as follows:

$$97 = 3 \cdot \underline{29} + \underline{10},$$
$$\underline{29} = 2 \cdot \underline{10} + \underline{9},$$
$$\underline{10} = 1 \cdot \underline{9} + \underline{1},$$
$$\underline{9} = 9 \cdot \underline{1}.$$

The numbers x_j are underlined to distinguish them from the quotients q_j. In this case we have

$$(x_0, x_1, x_2, x_3, x_4) = (97, 29, 10, 9, 1)$$

and

$$(q_1, q_2, q_3, q_4) = (3, 2, 1, 9). \tag{2.30}$$

We also have from the above equations that

$$\frac{97}{29} = 3 + \frac{10}{29},$$
$$\frac{29}{10} = 2 + \frac{9}{10},$$
$$\frac{10}{9} = 1 + \frac{1}{9}.$$

We can combine the first two of these to obtain

$$\frac{97}{29} = 3 + \cfrac{1}{2 + \frac{9}{10}},$$

and then use the third of the above equations to obtain

$$\frac{97}{29} = 3 + \cfrac{1}{2 + \cfrac{1}{1 + \frac{1}{9}}}. \tag{2.31}$$

Note how simply the above expression for the fraction $97/29$ is constructed from the quotients in (2.30). ■

The expression on the right of (2.31) is called a *simple continued fraction*. Since the form (2.31) is rather inconvenient to write, we usually express it in the condensed form

$$\frac{97}{29} = 3 + \frac{1}{2+} \; \frac{1}{1+} \; \frac{1}{9},$$

or even more simply as

$$\frac{97}{29} = [3, 2, 1, 9].$$

If we apply the process that begins with (2.26) and (2.27) to any numbers $x_0 > x_1 > 0$ such that x_0/x_1 is rational, we obtain

$$\frac{x_0}{x_1} = q_1 + \frac{1}{q_2+} \frac{1}{q_3+} \cdots \frac{1}{q_k}, \tag{2.32}$$

which we will also write as

$$\frac{x_0}{x_1} = [q_1, q_2, \ldots, q_k]. \tag{2.33}$$

If x_0/x_1 is irrational, we obtain the infinite continued fraction

$$\frac{x_0}{x_1} = q_1 + \frac{1}{q_2+} \frac{1}{q_3+} \cdots, \tag{2.34}$$

which we will alternatively write as

$$\frac{x_0}{x_1} = [q_1, q_2, q_3, \ldots]. \tag{2.35}$$

For any infinite continued fraction, we call $[q_1, q_2, \ldots, q_r]$, for any $r \geq 1$, its rth convergent. For the finite (terminating) continued fraction in (2.33), the rth convergent is defined for $1 \leq r \leq k$. It can be proved that the convergents of an infinite continued fraction do indeed, as the nomenclature implies, *converge* to the irrational number x_0/x_1. This means that we can obtain as close an approximation to x_0/x_1 as we wish by evaluating rth convergents for values of r that are sufficiently large.

Example 2.3.2 My pocket calculator has a key marked $1/x$. This is very helpful for computing continued fractions. For example, suppose we want to find the continued fraction for $\sqrt{5}$. My calculator also has a key marked $\sqrt{x}$. I insert the number 5 and press the $\sqrt{x}$ key, to obtain

$$2.236067977.$$

This gives $q_1 = 2$. I subtract 2 from the above number, leaving the fraction 0.236067977, and press the $1/x$ key to obtain

$$4.236067977.$$

This is very good news! For we see that $q_2 = 4$, and when we subtract 4 from 4.236067977, we again obtain the fraction 0.236067977. I don't need to press any more keys, for we can see that

$$\sqrt{5} = 2 + \frac{1}{4+} \frac{1}{4+} \frac{1}{4+} \cdots.$$

A little calculation gives the first few convergents,

$$\frac{2}{1}, \ \frac{9}{4}, \ \frac{38}{17}, \ \frac{161}{72}, \ \frac{682}{305}, \ \cdots$$

In decimal form, these are

$$2, \ 2.25, \ 2.2353, \ 2.23611, \ 2.236066, \ \ldots,$$

where the first two are given exactly, and the others to an appropriate number of digits to show how closely they approximate $\sqrt{5} \approx 2.236068$. If I begin with π instead of $\sqrt{5}$, I can repeatedly subtract the integer part and press the $1/x$ key to give a continued fraction that begins

$$\pi = 3 + \frac{1}{7+} \ \frac{1}{15+} \ \frac{1}{1+} \ \frac{1}{292+} \cdots.$$

This is a continued fraction that has no discernible pattern. Its first few convergents are

$$\frac{3}{1}, \ \frac{22}{7}, \ \frac{333}{106}, \ \frac{355}{113}, \ \frac{103993}{33102}.$$

The convergent $355/113$ approximates π with an error that is less than 0.0000003. This approximation was obtained by Zǔ Chōngzhī (429–500), a remarkable achievement that is one of the highlights of early Chinese mathematics. ∎

Although the first five convergents for π given in Example 2.3.2 are all correct, I would get wrong answers at some stage if I continued to compute further convergents with my pocket calculator. This is because my calculator works to ten decimal places, and the accuracy of the calculations is affected by rounding errors. This quite properly prompts the thought that we did not *prove* in Example 2.3.2 that

$$\sqrt{5} = [2, 4, 4, 4, \ldots].$$

However, we can now do *exactly* by algebraic means what we did above, but only approximately, with the aid of a calculator. Let us write

$$\sqrt{5} = 2 + \left(\sqrt{5} - 2\right),$$

where $\sqrt{5} - 2$ is the fractional part of $\sqrt{5}$. We now need to invert this fraction, and we find that

$$\sqrt{5} - 2 = \frac{\left(\sqrt{5} - 2\right)\left(\sqrt{5} + 2\right)}{\left(\sqrt{5} + 2\right)} = \frac{1}{\sqrt{5} + 2}, \tag{2.36}$$

since

$$\left(\sqrt{5} - 2\right)\left(\sqrt{5} + 2\right) = \left(\sqrt{5}\right)^2 - 2^2 = 5 - 4 = 1.$$

It follows from (2.36) that

$$\sqrt{5} = 2 + \frac{1}{\sqrt{5}+2} = \left[2, \sqrt{5}+2\right], \qquad (2.37)$$

and we similarly have

$$\sqrt{5} + 2 = 4 + \frac{1}{\sqrt{5}+2} = \left[4, \sqrt{5}+2\right]. \qquad (2.38)$$

Then the repeated application of (2.38) shows that

$$\sqrt{5} + 2 = [4, 4, 4, \dots],$$

and on combining this with (2.37), we have a *proof* that

$$\sqrt{5} = [2, 4, 4, 4, \dots].$$

Consider any convergent $[q_1, q_2, \dots, q_k]$. It follows from (2.32) and (2.33) that

$$[q_1, q_2, \dots, q_k] = q_1 + \frac{1}{[q_2, a_3, \dots, q_k]} = \left[q_1, [q_2, q_3, \dots, q_k]\right].$$

More generally, we have

$$[q_1, q_2, \dots, q_k] = \left[q_1, q_2, \dots, q_j, [q_{j+1}, q_{j+2}, \dots, q_k]\right], \qquad (2.39)$$

for $1 \leq j < k$. Note that the left side of (2.39) is a continued fraction with k parameters, and the right side is a continued fraction with the $j + 1$ parameters $q_1, q_2, \dots, q_j$, and $[q_{j+1}, q_{j+2}, \dots, q_k]$. We can use (2.39) repeatedly to evaluate any convergent "from the bottom up," via a sequence of calculations that begins

$$[q_1, q_2, \dots, q_k] = \left[q_1, \dots, q_{k-2}, [q_{k-1}, q_k]\right] = \left[q_1, \dots, q_{k-3}, [q_{k-2}, q_{k-1}, q_k]\right].$$

Each step of the calculation reduces the number of parameters in the continued fraction by one until, after $k - 2$ steps, we obtain

$$[q_1, q_2, \dots, q_k] = \left[q_1, [q_2, q_3, \dots, q_k]\right].$$

The above is just a formal account of the process that you may have used to check my calculations in Example 2.3.2, when I expressed the first five convergents of $\sqrt{5}$ as rational numbers.

Let us define two sequences (c_k) and (d_k), beginning with

$$c_1 = q_1, \; c_2 = q_2 q_1 + 1 \quad \text{and} \quad d_1 = 1, \; d_2 = q_2, \qquad (2.40)$$

and continuing by computing

$$c_{k+1} = q_k c_k + c_{k-1} \quad \text{and} \quad d_{k+1} = q_k d_k + d_{k-1}, \qquad (2.41)$$

for $k \geq 2$. Then it can be shown that

$$[q_1, q_2, \ldots, q_k] = \frac{c_k}{d_k} \qquad (2.42)$$

for all values of k for which the convergent $[q_1, q_2, \ldots, q_k]$ is defined. We can see immediately from (2.40) that

$$\frac{c_1}{d_1} = q_1 \quad \text{and} \quad \frac{c_2}{d_2} = \frac{q_2 q_1 + 1}{q_2} = [q_1 q_2],$$

and so the kth convergent is indeed given by c_k/d_k for $k = 1$ and $k = 2$. See, for example, Hardy and Wright [12] or Phillips [14] for a proof that (2.42) holds for all k. This gives a "top down" method of evaluating the convergents of a continued fraction, given that we know the values of the numbers $q_1, q_2, \ldots$, that is much superior to the "bottom up" method described above.

Let us look at the difference between consecutive convergents of a continued fraction. We have

$$\frac{c_k}{d_k} - \frac{c_{k-1}}{d_{k-1}} = \frac{c_k d_{k-1} - c_{k-1} d_k}{d_k d_{k-1}}, \qquad (2.43)$$

and let us examine further the numerator of the last fraction, since we expect it to be small compared with the denominator, $d_k d_{k-1}$, when k is large. We can use (2.41), with k replaced by $k - 1$, to obtain

$$c_k d_{k-1} - c_{k-1} d_k = (q_{k-1} c_{k-1} + c_{k-2}) d_{k-1} - c_{k-1}(q_{k-1} d_{k-1} + d_{k-2}),$$

and thus

$$c_k d_{k-1} - c_{k-1} d_k = -(c_{k-1} d_{k-2} - c_{k-2} d_{k-1}). \qquad (2.44)$$

This shows that as k varies, the numbers $c_k d_{k-1} - c_{k-1} d_k$ are all equal, apart from the sign, and we see from (2.40) that

$$c_2 d_1 - c_1 d_2 = (q_2 q_1 + 1) - q_1 q_2 = 1. \qquad (2.45)$$

Thus $c_k d_{k-1} - c_{k-1} d_k$ takes the values $+1$, when k is even, and -1, when k is odd, and so we can write

$$c_k d_{k-1} - c_{k-1} d_k = (-1)^k, \quad k \geq 2. \qquad (2.46)$$

It then follows from (2.43) that

$$\frac{c_k}{d_k} - \frac{c_{k-1}}{d_{k-1}} = \frac{(-1)^k}{d_k d_{k-1}}. \qquad (2.47)$$

Since we can see from the recurrence relation that $d_k \to \infty$ as $k \to \infty$, this shows that

$$\frac{c_k}{d_k} - \frac{c_{k-1}}{d_{k-1}} \to 0 \quad \text{as} \quad k \to \infty.$$

We need to do a little more work, but this is the beginning of an argument that leads to a proof that the convergents do indeed converge to the value of the continued fraction.

Example 2.3.3 Using the values $q_1 = 2$ and $q_j = 4$ for all $j \geq 2$, let us compute the first few convergents of the simple continued fraction for $\sqrt{5}$, which we discussed in Example 2.3.2. The first few values of the sequences (c_k) and (d_k) are given in Table 2.6. Although it just beyond the limits of my pocket calculator to determine the accuracy of the ninth convergent, it differs from $\sqrt{5}$ by less than 3 units in the 11th place after the decimal point. ∎

k	1	2	3	4	5	6	7	8	9
c_k	2	9	38	161	682	2889	12238	51841	219602
d_k	1	4	17	72	305	1292	5473	23184	98209

TABLE 2.6. The convergents c_k/d_k for the continued fraction for $\sqrt{5}$.

Obviously, the first convergent, c_1/d_1, is smaller than the value of the continued fraction. We can now show that the odd-order convergents, c_1/d_1, $c_3/d_3, \ldots$, form an increasing sequence, and the even-order convergents, c_2/d_2, $c_4/d_4, \ldots$, form a decreasing sequence. For we have, again using the recurrence relations in (2.41),

$$c_k d_{k-2} - c_{k-2} d_k = (q_{k-1} c_{k-1} + c_{k-2}) d_{k-2} - c_{k-2}(q_{k-1} d_{k-1} + d_{k-2}),$$

so that

$$c_k d_{k-2} - c_{k-2} d_k = q_{k-1}(c_{k-1} d_{k-2} - c_{k-2} d_{k-1}),$$

and we see from (2.46) that

$$c_k d_{k-2} - c_{k-2} d_k = (-1)^{k-1} q_{k-1}.$$

Finally, since

$$\frac{c_k}{d_k} - \frac{c_{k-2}}{d_{k-2}} = \frac{c_k d_{k-2} - c_{k-2} d_k}{d_k d_{k-2}},$$

we obtain

$$\frac{c_k}{d_k} - \frac{c_{k-2}}{d_{k-2}} = \frac{(-1)^{k-1} q_{k-1}}{d_k d_{k-2}}. \tag{2.48}$$

If k is odd, the factor $(-1)^{k-1}$ has the value $+1$, and so the right side of (2.48) is positive. Thus we see from (2.48) that the odd-order convergents form an increasing sequence. Assuming convergence, this sequence approaches the value of the continued fraction from below. If k is even, the factor $(-1)^{k-1}$ has the value -1, and we see that the even-order convergents form a decreasing sequence. Again assuming convergence, the even-order sequence approaches the value of the continued fraction from above. The value of the continued fraction lies between any pair of consecutive convergents, and so lies within an interval that, as we see from (2.47), diminishes as k is increased. This observation, which is equivalent to saying that each convergent c_k/d_k is closer to the value of the continued fraction than it is to the next convergent c_{k+1}/d_{k+1}, leads us to a very simple estimate of the error at any stage. Thus, from (2.47) with k replaced by $k+1$, the difference between c_k/d_k and the value of the continued fraction is less than $1/(d_{k+1}d_k)$.

Consider an infinite continued fraction of the form

$$\alpha = [x, \bar{y}] = [x, y, y, y, \ldots], \qquad (2.49)$$

where we have written $\bar{y}$ to denote the value y repeated indefinitely, and

$$x = [q_1, q_2, \ldots, q_m], \quad y = [q_{m+1}, q_{m+2}, \ldots, q_{m+n}]. \qquad (2.50)$$

This is called a *periodic* continued fraction. For example, the continued fraction $\sqrt{5} = [2, \bar{4}] = [2, 4, 4, 4, \ldots]$, given above, is periodic, as is the infinite continued fraction $[4, 2, \overline{1, 2, 3}] = [4, 2, 1, 2, 3, 1, 2, 3, 1, 2, 3, \ldots]$.

To evaluate the infinite continued fraction α in (2.49), it is convenient to replace $\bar{y}$ by z and write

$$z = [y, y, y, \ldots] = [y, z] = y + \frac{1}{z}, \qquad (2.51)$$

so that

$$z^2 = yz + 1.$$

On completing the square, as we did with equation (1.52), we find that

$$\left(z - \frac{1}{2}y\right)^2 = 1 + \frac{1}{4}y^2$$

and hence obtain

$$z - \frac{1}{2}y = \pm\left(1 + \frac{1}{4}y^2\right)^{1/2}, \qquad (2.52)$$

where the number on the right of (2.52) denotes plus or minus the square root of $1 + \frac{1}{4}y^2$. It follows from (2.52) that z must have the value $z = z_1$ or $z = z_2$, say, where

$$z_1 = \frac{1}{2}y + \left(1 + \frac{1}{4}y^2\right)^{1/2} \quad \text{and} \quad z_2 = \frac{1}{2}y - \left(1 + \frac{1}{4}y^2\right)^{1/2}.$$

Since

$$\left(1 + \frac{1}{4}y^2\right)^{1/2} > \left(\frac{1}{4}y^2\right)^{1/2} = \frac{1}{2}y,$$

we see that z_2 is negative. Since z has to be positive, we conclude that

$$z = z_1 = \frac{1}{2}y + \left(1 + \frac{1}{4}y^2\right)^{1/2}. \qquad (2.53)$$

We obtain from (2.49) and (2.51) that

$$\alpha = [x, y, y, y, \ldots] = [x, z] = x + \frac{1}{z}. \tag{2.54}$$

Then, on writing $c = \frac{1}{2}y$ in (2.53), we may express z as

$$z = \left(1 + c^2\right)^{1/2} + c = \frac{\left(\left(1 + c^2\right)^{1/2} + c\right)\left(\left(1 + c^2\right)^{1/2} - c\right)}{\left(1 + c^2\right)^{1/2} - c},$$

and since

$$\left(\left(1 + c^2\right)^{1/2} + c\right)\left(\left(1 + c^2\right)^{1/2} - c\right) = \left(1 + c^2\right) - c^2 = 1,$$

we find that

$$z = \frac{1}{\left(1 + c^2\right)^{1/2} - c}.$$

We then obtain from (2.54) that

$$\alpha = x - \frac{1}{2}y + \left(1 + \frac{1}{4}y^2\right)^{1/2}. \tag{2.55}$$

It remains only to evaluate the rational numbers x and y, defined in (2.50), to obtain α from (2.55). For the special case $x = 2$ and $y = 4$, (2.55) gives $\alpha = \sqrt{5}$, which is consistent with our findings in Example 2.3.2.

Mathematicians have been interested in continued fractions since at least the time of Euclid, circa 300 BC. Although I have discussed only *simple* continued fractions above, there are also continued fractions in which the partial numerators are not restricted to have the value 1. For example, the infinite continued fraction

$$\sqrt{13} = 3 + \frac{4}{6+} \frac{4}{6+} \cdots$$

was first derived by Rafaello Bombelli (1526–1573). In the following century, William, Viscount Brouncker (1620–1684), who was the first president of the Royal Society of London, derived the continued fraction

$$\frac{4}{\pi} = 1 + \frac{1^2}{2+} \frac{3^2}{2+} \frac{5^2}{2+} \frac{7^2}{2+} \cdots,$$

taking as his starting point the infinite product

$$\frac{\pi}{2} = \frac{2 \cdot 2 \cdot 4 \cdot 4 \cdot 6 \cdot 6 \cdot 8 \cdot 8}{1 \cdot 3 \cdot 3 \cdot 5 \cdot 5 \cdot 7 \cdot 7 \cdot 9} \cdots,$$

which was obtained by John Wallis (1616–1703). Brouncker's continued fraction involving π has a pleasingly regular pattern to it, a property that is not shared by the simple continued fraction for π that is given above, in Example 2.3.2. There are also continued fractions involving the famous number e, obtained by Leonhard Euler (1707–1783). We can define e as the sum of the infinite series

$$e = 1 + \frac{1}{1!} + \frac{1}{2!} + \frac{1}{3!} + \frac{1}{4!} + \cdots, \tag{2.56}$$

where $n! = 1 \times 2 \times 3 \times \cdots \times n$ is called n factorial. The first nine terms of the above infinite series suffice to determine that the value of e is approximately 2.71828. Euler derived the infinite continued fractions

$$e = 1 + [1, 1, 2, 1, 1, 4, 1, 1, 6, 1, 1, 8, \ldots],$$

$$\frac{e-1}{e+1} = [0, 2, 6, 10, 14, 18, \ldots], \tag{2.57}$$

and

$$\frac{1}{2}(e - 1) = [0, 1, 6, 10, 14, 18, \ldots]. \tag{2.58}$$

We can deduce (2.57) from (2.58). (See Problem 2.3.7.) It is generally believed that Euler was the first to establish that the number e is irrational.

Problem 2.3.1 Verify that

$$\frac{29}{18} = [1, 1, 1, 1, 1, 3] \quad \text{and} \quad \frac{21}{13} = [1, 1, 1, 1, 1, 2].$$

Problem 2.3.2 Show that

$$\sqrt{3} = [1, \overline{1, 2}].$$

Problem 2.3.3 Find the simple continued fraction for $\sqrt{2}$.

Problem 2.3.4 Write $\alpha = \sqrt{n^2 + 1}$, where n is any positive integer, and verify that $(\alpha + n)(\alpha - n) = 1$. Hence show that $\alpha = [n, \alpha + n]$, and also verify that

$$\alpha + n = [2n, \alpha + n] = [\overline{2n}].$$

Deduce that $\sqrt{n^2 + 1} = [n, \overline{2n}]$. Hence write down continued fractions for $\sqrt{2}$, $\sqrt{5}$, $\sqrt{10}$, and $\sqrt{17}$.

Problem 2.3.5 Verify that $\sqrt{n^2 + 2} = [n, \overline{n, 2n}]$ for every integer $n \geq 1$, and so write down continued fractions for $\sqrt{3}$, $\sqrt{6}$, $\sqrt{11}$, and $\sqrt{18}$.

Problem 2.3.6 Show that

$$\sqrt{n(n + 2)} = [n, \overline{1, 2n}], \qquad n \geq 1,$$

and

$$\sqrt{n^2 + 2n - 1} = [n, \overline{1, n - 1, 1, 2n}], \qquad n \geq 2,$$

and use these results to write down the continued fractions for $\sqrt{3}$, $\sqrt{7}$, $\sqrt{8}$, $\sqrt{14}$, and $\sqrt{15}$.

Problem 2.3.7 Let

$$\alpha = \frac{e-1}{e+1} \quad \text{and} \quad \beta = \frac{1}{2}(e-1).$$

Show that $1/\alpha = 1 + 1/\beta$, and so deduce (2.57) from (2.58).

2.4   The Euclidean Algorithm

We defined the division algorithm in Section 2.1 and applied it to express a given positive integer in a given base b. Then in Section 2.3 we used the division algorithm to develop the theory of continued fractions. In this section we see how the division algorithm leads us to some interesting and important results in the theory of numbers. Let us begin with any two positive integers n_0 and n_1, with $n_0 > n_1$, and apply the division algorithm repeatedly, as we did with the real numbers x_0 and x_1, beginning with (2.26). We obtain the equations

$$
\begin{aligned}
n_0 &= q_1 n_1 + n_2, \\
n_1 &= q_2 n_2 + n_3, \\
&\vdots \\
n_{k-2} &= q_{k-1} n_{k-1} + n_k, \\
n_{k-1} &= q_k n_k.
\end{aligned}
\tag{2.59}
$$

This process, which appears in the form of Proposition 2 in Book VII of Euclid's *Elements*, is called the Euclidean algorithm. We will see that the last number in the sequence $n_0, n_1, \ldots, n_k$ is the *greatest common divisor* of n_0 and n_1, which we will denote by $\gcd(n_0, n_1)$. As the name suggests, the greatest common divisor of two positive integers is the largest integer that divides both numbers. For example, $\gcd(18, 12) = 6$.

Example 2.4.1 Let us apply the Euclidean algorithm to the positive integers $n_0 = 101200$ and $n_1 = 84042$. We obtain

$$
\begin{aligned}
\underline{101200} &= 1 \cdot \underline{84042} + \underline{17158}, \\
\underline{84042} &= 4 \cdot \underline{17158} + \underline{15410}, \\
\underline{17158} &= 1 \cdot \underline{15410} + \underline{1748}, \\
\underline{15410} &= 8 \cdot \underline{1748} + \underline{1426}, \\
\underline{1748} &= 1 \cdot \underline{1426} + \underline{322}, \\
\underline{1426} &= 4 \cdot \underline{322} + \underline{138}, \\
\underline{322} &= 2 \cdot \underline{138} + \underline{46}, \\
\underline{138} &= 3 \cdot \underline{46}.
\end{aligned}
$$

If we factorize 101200 and 84042, we obtain

$$101200 = 2^4 \cdot 5^2 \cdot 11 \cdot 23 \quad \text{and} \quad 84042 = 2 \cdot 3^2 \cdot 7 \cdot 23 \cdot 29.$$

We can see directly from the above factorizations that

$$\gcd(101200, 84042) = 2 \cdot 23 = 46. \qquad \blacksquare$$

We see from the first equation in (2.59) that $\gcd(n_0, n_1) \mid n_2$, since it is clear that any positive integer that divides n_0 and n_1 must divide n_2. From the definition of the greatest common divisor we also have $\gcd(n_0, n_1) \mid n_1$, and thus $\gcd(n_0, n_1)$ divides both n_1 and n_2. Since $\gcd(n_1, n_2)$ is the largest number that divides n_1 and n_2, it follows that

$$\gcd(n_0, n_1) \mid \gcd(n_1, n_2). \tag{2.60}$$

If we begin again with the equation $n_0 = q_1 n_1 + n_2$, we can see that $\gcd(n_1, n_2)$ divides n_0 and argue similarly that since $\gcd(n_1, n_2)$ also divides n_1, we have

$$\gcd(n_1, n_2) \mid \gcd(n_0, n_1). \tag{2.61}$$

We deduce from (2.60) and (2.61) that

$$\gcd(n_0, n_1) = \gcd(n_1, n_2).$$

As we work through the equations (2.59) created by the Euclidean algorithm, we find similarly that

$$\gcd(n_0, n_1) = \gcd(n_1, n_2) = \cdots = \gcd(n_{k-1}, n_k),$$

and the final equation of (2.59) shows that $\gcd(n_{k-1}, n_k) = n_k$. We have thus proved the following result.

Theorem 2.4.1 Let the Euclidean algorithm be applied to the two positive integers $n_0 > n_1$ to create the sequence of equations (2.59), where all but the last of these equations connect three consecutive members of the decreasing sequence of positive integers

$$n_0 > n_1 > n_2 > \cdots > n_{k-1} > n_k,$$

and the last equation is $n_{k-1} = q_k n_k$. Then the final number n_k is the greatest common divisor of the two initial numbers n_0 and n_1. $\qquad \blacksquare$

Example 2.4.2 Consider the k equations (2.59) that define how the integers $n_2, n_3, \ldots, n_k$ are derived from the two initial integers n_0 and n_1, using the Euclidean algorithm. Let us ask the following question. For a given value of k, what is the smallest number n_0 that can occur in the first of the k equations? We find the answer by working back from the kth equation, making each number as small as possible. The smallest possible value for n_k is 1, and we cannot have $q_k = 1$, for this would yield the value 1 for n_{k-1}, and we must have $n_{k-1} > n_k$. Thus we must choose $q_k = 2$, and hence obtain $n_{k-1} = 2$. Pursuing our goal of making each number as small as possible, we now choose $q_{k-1} = q_{k-2} = \cdots = q_1 = 1$, and this completely determines the remaining numbers $n_{k-2}, n_{k-3}, \ldots, n_0$. For example, with $k = 7$, we obtain the equations

$$
\begin{aligned}
\underline{34} &= 1 \cdot \underline{21} + \underline{13}, \\
\underline{21} &= 1 \cdot \underline{13} + \underline{8}, \\
\underline{13} &= 1 \cdot \underline{8} + \underline{5}, \\
\underline{8} &= 1 \cdot \underline{5} + \underline{3}, \\
\underline{5} &= 1 \cdot \underline{3} + \underline{2}, \\
\underline{3} &= 1 \cdot \underline{2} + \underline{1}, \\
\underline{2} &= 2 \cdot \underline{1} ,
\end{aligned}
$$

in which the underlined numbers $n_k, n_{k-1}, \ldots, n_1, n_0$ are

$$1, \quad 2, \quad 3, \quad 5, \quad 8, \quad 13, \quad 21, \quad 34.$$

These are members of the *Fibonacci* sequence, and we have

$$\frac{34}{21} = [1, 1, 1, 1, 1, 1, 2] = [1, 1, 1, 1, 1, 1, 1, 1], \tag{2.62}$$

which is the eighth convergent of the infinite continued fraction $[1, 1, 1, \ldots]$. The ratio of any two consecutive Fibonacci numbers is a convergent of this infinite continued fraction. ∎

Let us now consider another application of the Euclidean algorithm. Given the equation

$$ax + by = c, \tag{2.63}$$

where a, b, and c are positive integers, can we find integers x and y that satisfy this Diophantine equation?

For any choice of integers x and y, $\gcd(a, b) \mid (ax + by)$, and so for any integers x and y that satisfy (2.63), we must have $\gcd(a, b) \mid c$. Thus, if c is not a multiple of $\gcd(a, b)$, there is no solution of (2.63). For example, the Diophantine equation

$$2x + 6y = 5$$

has no solution. Let us therefore consider the case in which c *is* a multiple of $\gcd(a, b)$, and write

$$a = \gcd(a, b) \cdot a_1, \quad b = \gcd(a, b) \cdot b_1, \quad c = \gcd(a, b) \cdot c_1. \tag{2.64}$$

Since $\gcd(a, b)$ is the greatest common divisor of a and b, we see from (2.64) that a_1 and b_1 can have no common factor other than 1, so that

$\gcd(a_1, b_1) = 1$. We now substitute the values of a, b, and c given in (2.64) into equation (2.63), and divide out by the common factor $\gcd(a, b)$, to obtain the equation

$$a_1 x + b_1 y = c_1, \quad \text{where} \quad \gcd(a_1, b_1) = 1. \tag{2.65}$$

If we now define x_1 and y_1 from

$$x = c_1 x_1 \quad \text{and} \quad y = c_1 y_1,$$

substitute these values of x and y into equation (2.65), and divide out by the common factor c_1, we obtain the equation

$$a_1 x_1 + b_1 y_1 = 1, \quad \text{where} \quad \gcd(a_1, b_1) = 1. \tag{2.66}$$

We will now use an argument based on the Euclidean algorithm to show that every Diophantine equation of the form (2.66) is satisfied by an infinite number of pairs of values x_1 and y_1. An example will show us how to obtain one solution of an equation of the form (2.66).

Example 2.4.3 Let us consider the Diophantine equation

$$53x_1 + 12y_1 = 1. \tag{2.67}$$

We note that $\gcd(53, 12) = 1$. Beginning with the numbers $n_0 = 53$ and $n_1 = 12$, we apply the Euclidean algorithm to obtain

$$\underline{53} = 4 \cdot \underline{12} + \underline{5}, \tag{2.68}$$
$$\underline{12} = 2 \cdot \underline{5} + \underline{2}, \tag{2.69}$$
$$\underline{5} = 2 \cdot \underline{2} + \underline{1}. \tag{2.70}$$

Since equation (2.70) ends with $\underline{1}$, this confirms that $\gcd(53, 12) = 1$. We rearrange (2.70) to obtain

$$\underline{1} = \underline{5} - 2 \cdot \underline{2},$$

and substitute for $\underline{2}$, using (2.69), to obtain

$$\underline{1} = \underline{5} - 2 \cdot (\underline{12} - 2 \cdot \underline{5}) = 5 \cdot \underline{5} - 2 \cdot \underline{12}.$$

We now substitute for $\underline{5}$, using equation (2.68), to obtain

$$\underline{1} = 5 \cdot (\underline{53} - 4 \cdot \underline{12}) - 2 \cdot \underline{12} = 5 \cdot \underline{53} - 22 \cdot \underline{12}.$$

Thus we have shown that

$$5 \cdot \underline{53} - 22 \cdot \underline{12} = \underline{1},$$

so that $x_1 = 5$ and $y_1 = -22$ is a solution of (2.67). ■

We can always find *one* solution of an equation of the form (2.66) by following the method that we used in Example 2.4.3. For we can execute the Euclidean algorithm based on the two numbers $n_0 = a_1$ and $n_1 = b_1$, assuming that $a_1 > b_1$. This ends after, say, k stages, with $n_k = 1$, since $\gcd(a_1, b_1) = 1$. Then, beginning with the kth equation, we work back through the equations, as we did in Example 2.4.3, to obtain numbers $\overline{x}$ and $\overline{y}$ such that

$$a_1\overline{x} + b_1\overline{y} = 1.$$

If we now choose

$$x_1 = \overline{x} - b_1 t, \qquad y_1 = \overline{y} + a_1 t, \tag{2.71}$$

where t is any integer, then

$$a_1 x_1 + b_1 y_1 = a_1(\overline{x} - b_1 t) + b_1(\overline{y} + a_1 t) = a_1\overline{x} + b_1\overline{y} = 1,$$

showing that (2.71) gives an infinite number of solutions of the Diophantine equation (2.66). For example, (2.67) has the solutions

$$x_1 = 5 - 12t, \qquad y_1 = -22 + 53t,$$

where t is any integer. We can summarize our findings on the Diophantine equation $ax + by = c$ as follows.

Theorem 2.4.2 The Diophantine equation $ax + by = c$ has solutions if and only if $\gcd(a, b) \mid c$. If this condition holds, the above equation can be transformed into the equation

$$a_1 x_1 + b_1 y_1 = 1, \tag{2.72}$$

where the coefficients a_1 and b_1 are given by

$$a = \gcd(a, b) \cdot a_1, \qquad b = \gcd(a, b) \cdot b_1,$$

so that $\gcd(a_1, b_1) = 1$, and

$$\gcd(a, b) \cdot x = cx_1, \qquad \gcd(a, b) \cdot y = cy_1.$$

Then the transformed Diophantine equation (2.72) has the solutions

$$x_1 = \overline{x} - b_1 t, \qquad y_1 = \overline{y} + a_1 t,$$

for any integer t, where $x_1 = \overline{x}$, $y_1 = \overline{y}$ is the solution of (2.72) obtained by applying the Euclidean algorithm to a_1 and b_1 and working back through the equations generated by the algorithm. ∎

Another application of the Euclidean algorithm is the following result, which is of great importance in the theory of numbers.

Theorem 2.4.3 If a prime p divides the product ab, where a and b are positive integers, then p divides at least one of a and b.

Proof. Let us suppose that the prime p divides ab, and does not divide a. Then, as we showed above in discussing the solution of the Diophantine equation (2.66), since $\gcd(a, p) = 1$, there exist integers x and y such that

$$ax + py = 1,$$

and on multiplying throughout by b, we obtain

$$abx + pby = b. \tag{2.73}$$

Since p divides ab and p divides pb, we see that p divides the left side of equation (2.73), and so divides b. This completes the proof.   ■

Problem 2.4.1 Find $\gcd(16617, 6322)$ by applying the Euclidean algorithm.

Problem 2.4.2 Find an infinite number of pairs of integers x and y that satisfy the Diophantine equation

$$17x - 14y = 1.$$

Problem 2.4.3 Use the Euclidean algorithm to show that for any choice of positive integer n, the integers $12n+5$ and $3n+2$ have no common factor greater than 1. Hence find integers x and y that satisfy the equation

$$(12n + 5)x + (3n + 2)y = 1.$$

Problem 2.4.4 Show that if a, b, and c are positive integers, then

$$\gcd(a + cb, b) = \gcd(a, b).$$

Problem 2.4.5 Prove that $\gcd(a, b)$ is the smallest positive integer that can be written as a linear combination of a and b, that is, the smallest positive integer of the form $ax + by$, where x and y are integers.

Problem 2.4.6 How would you find the greatest common divisor of three positive integers a, b, and c?

2.5 Infinity

It seems plausible that mankind's appreciation of the concept of number began with the natural numbers, $1, 2, 3, \ldots$, and that at some early stage there emerged an understanding that the set of all natural numbers is infinite. Leopold Kronecker (1823–1891) recognized the very special status of these numbers when he said, "God made the natural numbers; all else

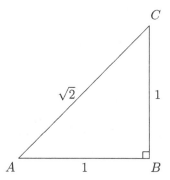

FIGURE 2.2. The right-angled triangle ABC with $|AB| = |BC| = 1$ has hypotenuse AC, whose length is the irrational number $\sqrt{2}$.

is the work of man." The Greeks were familiar with the positive rational numbers. These are of the form m/n, where m and n are natural numbers. Every *length* can be approximated as closely as we please by a positive rational number. However, it has been known to mathematicians since at least the time of Pythagoras, who lived in the sixth century BC, that not all lengths are rational numbers. One example is the length x of the hypotenuse (the longest side) of the right-angled triangle in which the sides that enclose the right angle are of unit length. (See Figure 2.2.) Thus, from Pythagoras's theorem, we have $x^2 = 1^2 + 1^2 = 2$. We write $x = \sqrt{2}$ to denote the positive solution of the equation $x^2 = 2$ and call x the square root of 2. We will now show that $\sqrt{2}$ is an irrational number.

Suppose now that we can write $x = \sqrt{2}$ in the form m/n, where m and n are positive integers. We can assume that we have canceled any common factor possessed by m and n, so that $(m, n) = 1$. Thus

$$x^2 = \frac{m^2}{n^2} = 2,$$

and so

$$m^2 = 2n^2. \tag{2.74}$$

This implies that m is even, and so $m = 2m_1$, say, where m_1 is a positive integer. This yields $4m_1^2 = 2n^2$, and hence

$$2m_1^2 = n^2. \tag{2.75}$$

We deduce from (2.75) that n is even. Since m and n are both even, they have a common factor, which contradicts our above statement that

$$x = \sqrt{2} = \frac{m}{n},$$

where m and n have no common factor. Thus $\sqrt{2}$ is *not* a rational number. In the above proof we make an *assumption* (that $\sqrt{2}$ is a rational number) and obtain a *contradiction*, causing us to conclude that the original hypothesis is untenable. This is called *proof by contradiction* or (in Latin) *reductio ad absurdum.*

The above proof is easily adapted to show that $\sqrt{p}$ is irrational where p is any prime number, that is, a positive integer that is not divisible by any positive integer other than 1 and itself. Another example of an irrational number (there is no scarcity of these!) is the number e, defined in (2.56). I will not prove here that e is irrational. Every irrational number can be approximated as closely as we please by a rational number. For example,

$$1.414 < \sqrt{2} < 1.415 \quad \text{and} \quad 2.718 < e < 2.719.$$

The set of all lengths, which is the union of the set of rational numbers and the set of irrational numbers, is called the set of *real* numbers, denoted by $\mathbb{R}$. We write $x \in \mathbb{R}$ to mean that x belongs to the set $\mathbb{R}$; that is, x is a real number. To help our understanding of the natural numbers, rational numbers, and irrational numbers, we require the following definition.

Definition 2.5.1 Consider two sets S_1 and S_2. If to each $x \in S_1$ there corresponds a unique $y \in S_2$, and to each $y \in S_2$ there corresponds a unique $x \in S_1$, we say that there is a *one-to-one correspondence* between the two sets. Thus, in a one-to-one correspondence, the elements of the two sets are paired off, and we say that the two sets are *equivalent* and that they have the same *cardinal number.* ∎

Remark 2.5.1 The word "equivalent" that occurs in Definition 2.5.1 is compatible with the mathematical concept of equivalence relation to which we refer in Remark 2.2.1. Thus the three properties of an equivalence relation apply:

1. A set S is equivalent to itself (reflexive property).

2. If a set S_1 is equivalent to S_2, then S_2 is equivalent to S_1 (symmetric property).

3. If S_1 is equivalent to S_2, and S_2 is equivalent to S_3, then S_1 is equivalent to S_3 (transitive property). ∎

Georg Cantor (1845–1918) applied the ideas embodied in Definition 2.5.1 to create some very exciting mathematics involving cardinal numbers. Of course, if a set has a finite number of elements, its cardinal number is simply the number of elements in the set. The story becomes much more interesting, and more puzzling, when we consider infinite sets. For example, there are "obviously" more positive integers (natural numbers) than there are even positive integers. However, these two sets have the same cardinal number! This follows from the one-to-one correspondence that is defined by identifying the positive integer n with the even positive integer $2n$. Cantor used the symbol $\aleph_0$ (the first letter of the Hebrew alphabet, *aleph*, with suffix 0) to denote the cardinal number of the natural numbers. Any set that is equivalent to the set of natural numbers has cardinal number $\aleph_0$. Such a set is called *denumerable*, which means countable. The cardinal number of *any* infinite set is said to be *transfinite*. We now state and prove a surprising result that was obtained by Cantor.

Theorem 2.5.1 The rational numbers have cardinal number $\aleph_0$, the same cardinal number as the natural numbers.

Proof. We need to show that there is a one-to-one correspondence between the *positive* rational numbers and the natural numbers. We begin by arranging the positive rational numbers in the following order:

$$\frac{1}{1}; \quad \frac{1}{2}, \frac{2}{1}; \quad \frac{1}{3}, \frac{2}{2}, \frac{3}{1}; \quad \frac{1}{4}, \frac{2}{3}, \frac{3}{2}, \frac{4}{1}; \quad \cdots . \tag{2.76}$$

Note that each "block" of numbers that lies between consecutive semicolons in (2.76) consists of positive rational numbers of the form m/n, where $m + n$ is a constant. We now go through the sequence of rational numbers set out in (2.76) and *delete* numbers that are equivalent to rational numbers that have already appeared in the sequence. We thus delete every rational number a/b where $(a, b) > 1$. The first number to be deleted is $\frac{2}{2}$, because it is equivalent to $\frac{1}{1}$. Having thus attained an ordering of the positive rational numbers, we identify $\frac{1}{1}$ with 1, $\frac{1}{2}$ with 2, $\frac{2}{1}$ with 3, $\frac{1}{3}$ with 4, $\frac{3}{1}$ with 5, and so on. Similarly, we can establish a one-to-one correspondence between the negative rational numbers and the natural numbers. Finally, if we begin with 0, and take positive and negative rational numbers alternately, we have a one-to-one correspondence between the set of *all* rational numbers and the natural numbers.   ∎

The set of rational numbers is the set of all distinct solutions of equations of the form

$$a_0 x + a_1 = 0, \tag{2.77}$$

where $a_0 \neq 0$ and a_1 are integers. Let us now define a generalization of the rational numbers.

Definition 2.5.2 The set of all complex numbers that are solutions of equations of the form

$$a_0 x^n + a_1 x^{n-1} + \cdots + a_{n-1} x + a_n = 0, \tag{2.78}$$

where $n \geq 1$ and $a_0, a_1, \ldots, a_n$ are integers, with a_0 nonzero, is called the set of *algebraic* numbers. A number that is not algebraic is called *transcendental*.   ∎

Although the set of algebraic numbers may seem very much "bigger" than the set of rational numbers, Cantor showed that both sets have cardinal number $\aleph_0$. In the proof of this result, we require the following definition.

Definition 2.5.3 For any real number t, we define

$$|t| = \begin{cases} t, & \text{if } t \geq 0, \\ -t, & \text{if } t < 0. \end{cases} \tag{2.79}$$

and call $|t|$ the *absolute value* of t or the *modulus* of t. Note that $|t| \geq 0$. Equivalently, we could write $|t| = \left(t^2\right)^{1/2}$, and so the definition of the modulus of a real number is compatible with that of a complex number. (See Definition 1.5.3.)   ■

Theorem 2.5.2 The set of all algebraic numbers is denumerable.

Proof. Let

$$p(x) = a_0 x^n + a_1 x^{n-1} + \cdots + a_{n-1} x + a_n, \tag{2.80}$$

where $n \geq 1$ and $a_0, a_1, \ldots, a_n$ are integers, with a_0 nonzero, and define

$$N(p) = n + |a_0| + |a_1| + \cdots + |a_n|. \tag{2.81}$$

We note that $N(p)$ is a positive integer, and observe that $N(p) \geq 2$. We will assume, without proof, that the equation $p(x) = 0$ is satisfied by no more than n algebraic numbers, where n is the degree of the polynomial p. We see that there is only a finite number of polynomials associated with a given positive integer N, as in (2.81), and the corresponding polynomial equations have a finite number of distinct algebraic numbers as solutions. We can therefore take $N = 2, 3, \ldots$ in turn and, in principle, obtain all the algebraic numbers, deleting any that we have already encountered. Thus the set of all algebraic numbers is denumerable and so has the same cardinal number as the natural numbers.   ■

After the shock of Cantor's Theorems 2.5.1 and 2.5.2, you might wonder whether all infinite sets are denumerable. It was Cantor who also settled this question, as described in Theorem 2.5.3 below. However, before we study this result, let us observe that any real number that can be written as a *terminating* decimal can also be written as a nonterminating decimal. For example

$$\frac{1}{4} = 0.25 = 0.24999\ldots.$$

To give a unique decimal representation for every real number, let us agree that whenever we have such a choice we select the terminating form rather that the form with an infinite number of digits 9.

Theorem 2.5.3 The set of all real numbers x such that $0 < x < 1$ is *not* denumerable.

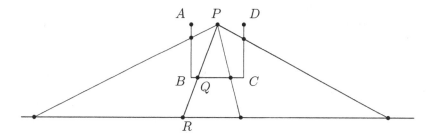

FIGURE 2.3. There is a one-to-one correspondence between the points in a line segment of finite length and the whole real line.

Proof. Suppose that there exists a one-to-one correspondence between the real numbers in $(0, 1)$ and the natural numbers. Then we could write down these real numbers in the order $r_1, r_2, r_3, \ldots$. Now let us construct a real number

$$x = 0.a_1 a_2 a_3 \ldots ,$$

where a_1, a_2, $a_3, \ldots$ are the decimal digits of x, and let us choose the nth digit, a_n, so that it is *different* from the nth digit of r_n, the nth real number in the above list. Thus, by construction, the real number x is different from every number in the whole list. This contradicts the assumption that there is a one-to-one correspondence between the real numbers in $[0, 1]$ and the natural numbers. ■

We can show that there is a one-to-one correspondence between the set of all x such that $0 < x < 1$ and the set of all positive real numbers. More generally, there is a one-to-one correspondence between the points in any finite line segment and the whole real line. Such a one-to-one correspondence is shown geometrically in Figure 2.3. The line segment AD is bent at intermediate points B and C. After bending the line segment we choose a point P midway between A and D and arrange that APD is parallel to the line shown at the bottom of the diagram, which depicts part of the real line. The one-to-one correspondence is determined by pairs of points, one on the bent segment $ABCD$, say Q, and one on the real line, say R, such that Q, R, and the fixed point P, lie on a straight line. Four pairs of points of this one-to-one correspondence are displayed in Figure 2.3. (Only one of the four pairs of points are labeled Q and R, for the sake of clarity.)

Definition 2.5.4 A set is said to be *nonempty* if it contains at least one element. Let S and S' denote two sets with the property that every element of S' also belongs to S. Then S' is said to be a *subset* of S, and we write $S' \subset S$. If, in addition, S' is nonempty and is not identical to S, we say that S' is a *proper* subset of S. ■

We cannot rely on our intuition when we are dealing with infinite sets. For example, the notion that the set of even positive integers is equivalent to the set of the natural numbers may seem rather strange at first, given that our intuition tells us that the first set appears to be half the *size* of the second. But this is the very point: What do we *mean* by "size"? Cantor's concept of the equivalence of sets gives us a means of comparing two sets and, in particular, two infinite sets. We have to accept the consequences of this approach, even if it means that a set can be equivalent to a proper subset of itself. Any infinite subset of the natural numbers is equivalent to the set of natural numbers, and so has cardinal number $\aleph_0$. Obvious examples are the square numbers, $1, 4, 9, \ldots$, the cubes, $1, 8, 27, \ldots$, or indeed any sequence of powers. Another infinite subset of the natural numbers is the set of primes, as we will see in Theorem 4.1.1.

There is an amusing tale, which is now part of mathematical folklore, concerning the mythical Hilbert's Hotel. This hotel, named after the mathematician David Hilbert (1862–1943), has an infinite number of rooms, numbered 1, 2, 3, and so on. (So we know that if all the real numbers wish to stay there, they can't have a room each! In fact, at least one room would have to contain an infinite number of real-number guests.) One evening a new guest arrived at Hilbert's Hotel and asked for a room. "One moment please," said the manager, and rushed off. On his return, he said "No problem. The hotel was full, but I moved the occupant of room 1 into room 2, the occupant of room 2 into room 3, and so on. Here is the key to room 1. I'm sorry, it doesn't have a sea view. Is that OK?" "Sure, that's fine," the guest replied, "I already feel bad about putting you to an infinite amount of trouble." Later, a bus arrived containing 20 would-be guests. The manager had a little difficulty in moving the guest from room 1 to room 21, since he had just arrived and was very tired and grumpy. In fact, all the original guests were grumpy at being moved for the second time. However, the manager moved the guest who was in room 2 to room 22, the one who was in room 3 to room 23, and so on, and so was able to accommodate the 20 new guests. An hour later, a Hilbert bus arrived. As you may have already guessed, it contained a denumerable number of new guests, numbered 1, 2, 3, and so on. I will leave you to think about how the manager was able to accommodate them all in single rooms. If you need a hint, think of the even numbers. (However, once you have thought about Problem 2.5.2 you will see that the manager could have avoided the above problems.)

We have seen that there are at least two infinite sets with different cardinal numbers, namely the set of natural numbers (or any set that is equivalent to it) and the set of real numbers (or any set equivalent to it). Is there any infinite set whose cardinal number is different from these two? This question was answered in the affirmative by Cantor himself, using the concept of subsets. We state and justify this substantial discovery of Cantor after the following definition and example.

Definition 2.5.5 Given any set S, which may be finite or infinite, we define a set T, called the *power set* of S, whose elements consist of all the subsets of S. The set T, which is often written as $\mathcal{P}(S)$, includes both the whole set S and the subset that contains no elements. We call this the latter subset the *empty set* and denote it by $\emptyset$. (You can think of the symbol $\emptyset$ as a zero with a slash through it.) ■

Example 2.5.1 Consider the set $S = \{a_1, a_2\}$. Its subsets are the empty set, the two subsets that contain just one element, and the whole set S. Thus we have

$$\mathcal{P}(S) = \Big\{\emptyset, \{a_1\}, \{a_2\}, \{a_1, a_2\}\Big\}.$$

Thus, for a set S that has 2 elements, the power set $\mathcal{P}(S)$ has 4 elements. More generally (see Problem 2.5.3), if a set S has n elements, the power set $\mathcal{P}(S)$ has 2^n elements. ■

Theorem 2.5.4 Let S denote any nonempty set S, which may be finite or infinite. Then the cardinal number of the power set $\mathcal{P}(S)$ is greater than the cardinal number of S.

Proof. First let us consider the case in which S is a nonempty set that has a finite number of elements, say n. Then the cardinal number of S is n, and the cardinal number of $\mathcal{P}(S)$ is 2^n. Since $2^n > n$ for all $n \geq 1$, this verifies the theorem when S is finite. It remains to consider the case in which S is an *infinite* set. Let us assume that $\mathcal{P}(S)$ is equivalent to S or to some subset of S and seek a contradiction. This implies that there exists a one-to-one correspondence between the elements of $\mathcal{P}(S)$ and S or a subset of S. To represent this one-to-one correspondence, let us write

$$x \longleftrightarrow S_x, \tag{2.82}$$

where x belongs to S or a subset of S, and S_x denotes the element of $\mathcal{P}(S)$ that is linked to x by the one-to-one correspondence. (Thus S_x is a subset of S.) We will now obtain a contradiction by finding an element of $\mathcal{P}(S)$ that is not linked to an element of S or a subset of S by the one-to-one correspondence.

Note that given any x in the correspondence denoted by (2.82), there are two possibilities: either x belongs to S_x or x does not belong to S_x. Let S^* denote the subset of S consisting of all elements x in the correspondence with the property that x is *not* contained in its corresponding set S_x. We now show that S^*, which is of course an element of $\mathcal{P}(S)$, cannot be any one of the elements S_x. For consider any S_x. If the corresponding x belongs to S_x, then x does not belong to S^*. On the other hand, if x does *not* belong to S_x, then x must belong to S^*. This shows that S^* differs from *every* S_x, and thus there is not a one-to-one correspondence between the elements of $\mathcal{P}(S)$ and S or a subset of S, and so the sets S and $\mathcal{P}(S)$ are not equivalent. However, we can set up an obvious one-to-one correspondence between S

and the subset of $\mathcal{P}(S)$ consisting of S itself. Thus the cardinal number of $\mathcal{P}(S)$ is greater than that of S, and this completes the proof.   ∎

Remark 2.5.2 Do not be too easily discouraged if you find the proof of Theorem 2.5.4 puzzling at a first reading. It is the most difficult proof in this book. However, it is worth persevering with the proof so that you can truly appreciate Cantor's genius and also *know* that the following amazing result is true.   ∎

Theorem 2.5.5 Let $\mathbb{N} = \mathbb{N}_1$ denote the set of natural numbers, and let us define the sets $\mathbb{N}_2$, $\mathbb{N}_3$, ... recursively by

$$\mathbb{N}_{k+1} = \mathcal{P}\left(\mathbb{N}_k\right), \quad \text{for} \quad k = 1, 2, 3, \ldots . \tag{2.83}$$

Then, corresponding to the infinite sequence $\mathbb{N}_1$, $\mathbb{N}_2$, $\mathbb{N}_3$, ..., there is an infinite sequence of increasing cardinal numbers.

Proof. The proof follows immediately from Cantor's Theorem 2.5.4.   ∎

Problem 2.5.1 Show that every transcendental number is irrational.

Problem 2.5.2 A denumerable number of Hilbert buses, each containing a denumerable number of would-be guests, arrive at Hilbert's Hotel, which is already full. How can all the guests be accommodated? Hint: begin by moving each current guest from room i to room 2^i and use rooms p_j^i with $j \geq 2$ for the new guests, where p_j denotes the jth prime.

Problem 2.5.3 Let S denote a set that has n elements, say $a_1, a_2, \ldots, a_n$. In choosing a subset of S, we have 2 choices of whether to include a_1 or not. Likewise, we have 2 choices of whether to include a_2 or not. Thus there are 2×2 choices concerning the inclusion of a_1 and a_2. Extend this argument to show that $\mathcal{P}(S)$ has 2^n elements. Note that the empty set corresponds to the choice where we exclude every element.

Problem 2.5.4 Prove by mathematical induction that $2^n > n$ for all positive integers n. Then argue that, since $2^n > 0$ for *all* integers n, the inequality $2^n > n$ holds for all integers.

3
Fibonacci Numbers

It is true that a mathematician who is not also something of a poet will never be a perfect mathematician.

Karl Weierstrass (1815–1897)

3.1   Recurrence Relations

In Section 2.4 we met the Fibonacci sequence (F_n), defined by the recurrence relation

$$F_{n+1} = F_n + F_{n-1}, \quad n \geq 1, \tag{3.1}$$

where $F_1 = F_2 = 1$. The first few Fibonacci numbers are

$$1, \quad 1, \quad 2, \quad 3, \quad 5, \quad 8, \quad 13, \quad 21, \quad 34, \quad 55, \quad 89, \quad 144, \quad 233.$$

This famous sequence is named after Leonardo of Pisa (circa 1175–1220), who is more usually called Fibonacci. The latter name is derived from Filius Bonaccii, meaning the son of Bonaccio. Fibonacci introduced his sequence in *Liber Abaci,* his "book of the abacus," which was published in 1202. The sequence (F_n) arose in the solution of a problem, discussed by Fibonacci in *Liber Abaci,* concerning the growth of a population of rabbits. Initially there is a single breeding pair of rabbits. Each pair of rabbits produces another breeding pair every month, and a new pair produces its first breeding pair offspring after two months. This is a very simple model, and there are no deaths. Fibonacci posed and subsequently answered the question of how many pairs of rabbits there are after one year.

Let r_j denote the number of pairs of rabbits existing at the beginning of month j. Then r_{j+2} is equal to the sum of the number of pairs alive at the beginning of month $j + 1$ plus the number of pairs born at the beginning of month $j + 2$, which equals the number of pairs alive two months earlier. Thus

$$r_{j+2} = r_{j+1} + r_j, \tag{3.2}$$

with $r_1 = r_2 = 1$, and so $r_j = F_j$ for all $j \geq 1$. Then the number of pairs existing after one year, that is, at the beginning of month 13, is $F_{13} = 233$.

We will see in the next section that as n increases, the Fibonacci number F_n grows like a multiple of x^n, where $x = \frac{1}{2}\left(\sqrt{5} + 1\right)$. Such a rate of growth is said to be *exponential*. Mathematicians find Fibonacci numbers fascinating because, as we will see, they have many interesting properties.

The Fibonacci sequence begins with $F_1 = 1$ and $F_2 = 1$. These are the *initial conditions* of the recurrence relation (3.1) that defines the rest of the sequence. It is clear from (3.1) that if we change the initial conditions, we change the sequence. Once we have chosen the initial conditions, the rest of the sequence is completely determined. Let us consider a sequence (S_n) defined by

$$S_{n+1} = S_n + S_{n-1}, \quad n \geq 2, \tag{3.3}$$

where S_1 and S_2 are still to be chosen. Obviously, each choice of initial conditions S_1 and S_2 determines a unique sequence (S_n) that satisfies (3.3), and the choice $S_1 = S_2 = 1$ gives the Fibonacci sequence. Let (S'_n) and (S''_n) denote any two sequences that satisfy (3.3), and write

$$S_n = a_1 S'_n + a_2 S''_n. \tag{3.4}$$

We say that S_n, defined by (3.4), is a *linear combination* of S'_n and S''_n. Using (3.4) and the fact that (S'_n) and (S''_n) satisfy (3.3), we have

$$S_{n+1} = a_1 S'_{n+1} + a_2 S''_{n+1} = a_1(S'_n + S'_{n-1}) + a_2(S''_n + S''_{n-1})$$
$$= (a_1 S'_n + a_2 S''_n) + (a_1 S'_{n-1} + a_2 S''_{n-1}) = S_n + S_{n-1},$$

and this shows that if two sequences satisfy the recurrence relation (3.3), any linear combination of these two sequences also satisfies (3.3).

As we saw in Example 2.4.2, the ratio of any two consecutive Fibonacci numbers is a convergent of the infinite continued fraction $[1, 1, 1, \ldots] = x$, say. Thus, for n large,

$$\frac{F_{n+1}}{F_n} \approx [1, 1, 1, \ldots] = x, \tag{3.5}$$

and this suggests that F_n behaves *approximately* like Cx^n for large n, where C is a positive constant and x is defined by (3.5). If we substitute Cx^n for

S_n in (3.3), we obtain

$$Cx^{n+1} = Cx^n + Cx^{n-1}, \quad n \geq 1.$$

On dividing by Cx^{n-1}, we see that x must satisfy the equation

$$x^2 = x + 1. \tag{3.6}$$

This is called the *characteristic equation* of the recurrence relation (3.3). We can complete the square and so express (3.6) in the form

$$\left(x - \frac{1}{2}\right)^2 = x^2 - x + \frac{1}{4} = \frac{5}{4},$$

so that

$$x - \frac{1}{2} = \pm\frac{\sqrt{5}}{2}.$$

We will denote these two solutions by α (*alpha*) and β (*beta*), the first two letters of the Greek alphabet. We have

$$\alpha = \frac{1}{2}\left(\sqrt{5}+1\right) \quad \text{and} \quad \beta = \frac{1}{2}\left(-\sqrt{5}+1\right). \tag{3.7}$$

Because β is negative, the solution that satisfies (3.5) is

$$\alpha = [1, 1, 1, \ldots] = \frac{1}{2}\left(\sqrt{5}+1\right). \tag{3.8}$$

Since $S_n = \alpha^n$ and $S_n = \beta^n$ both satisfy the recurrence relation (3.3), and we proved above that any linear combination of two solutions also satisfies (3.3), we see that

$$S_n = A\alpha^n + B\beta^n \tag{3.9}$$

satisfies (3.3). The quantity S_n, defined by (3.9), is called the *general solution* of the recurrence relation (3.3). We can now use the initial conditions, the values of S_1 and S_2, to determine the values of A and B. In particular, for the special case of the Fibonacci sequence we can write

$$F_n = A\alpha^n + B\beta^n, \tag{3.10}$$

and we seek values of A and B such that

$$F_1 = A\alpha + B\beta = 1 \quad \text{and} \quad F_2 = A\alpha^2 + B\beta^2 = 1. \tag{3.11}$$

Since α and β both satisfy (3.6), we see that

$$\alpha^2 - \alpha = \beta^2 - \beta = 1.$$

If we subtract the two equations in (3.11), we find that

$$A\left(\alpha^2 - \alpha\right) + B\left(\beta^2 - \beta\right) = 0,$$

and thus $A + B = 0$. We then see from the first of the equations in (3.11) that

$$A = -B = \frac{1}{\alpha - \beta}.$$

Thus the nth Fibonacci number is given by

$$F_n = \frac{\alpha^n - \beta^n}{\alpha - \beta}. \tag{3.12}$$

Note that although α and β are irrational numbers, the Fibonacci number F_n, defined by (3.12), is a positive integer. Although this explicit representation of F_n is called the Binet form, named after Jacques Binet (1786–1856), it was known much earlier to Abraham De Moivre.

Let us consider how rapidly F_n grows as n increases. Since

$$x^2 - x - 1 = (x - \alpha)(x - \beta) = x^2 - (\alpha + \beta)x + \alpha\beta,$$

we deduce that

$$\alpha + \beta = 1 \qquad \text{and} \qquad \alpha\beta = -1, \tag{3.13}$$

and we can see from (3.7) that

$$\alpha - \beta = \sqrt{5}. \tag{3.14}$$

The value of β is approximately -0.618, and thus $|\beta| < 1$. (Recall that $|\beta|$ denotes the absolute value of β, as given by Definition 2.5.3. Thus $|\beta|$ is approximately 0.618.) It follows that β^n tends to zero as n tends to infinity, and it is then clear from (3.12) that the growth of F_n is governed by α, rather than β. Since $\alpha - \beta = \sqrt{5}$, we see from (3.12) that F_n is given approximately by $\alpha^n/\sqrt{5}$. Since

$$|\beta^n| \leq |\beta|$$

for $n \geq 1$, this approximation has an error that satisfies

$$0 < \left| \frac{\beta^n}{\alpha - \beta} \right| \leq \left| \frac{\beta}{\alpha - \beta} \right| = \frac{\sqrt{5} - 1}{2\sqrt{5}} = \frac{1}{2} - \frac{1}{2\sqrt{5}} < \frac{1}{2},$$

and so F_n is the nearest integer to $\alpha^n/\sqrt{5}$. Since the error $\beta^n/(\alpha - \beta)$ alternates in sign with n, this estimate of F_n is alternately too large and too small, and the error tends to zero as n tends to infinity. For example,

$$F_{12} \approx \frac{\alpha^{12}}{\sqrt{5}} = 144.00139 \quad \text{and} \quad F_{13} \approx \frac{\alpha^{13}}{\sqrt{5}} = 232.99914,$$

to five decimal places, confirming that $F_{12} = 144$ and $F_{13} = 233$.

Let us look again at the ratio F_{n+1}/F_n. We see from (3.12) that

$$\frac{F_{n+1}}{F_n} = \frac{\alpha^{n+1} - \beta^{n+1}}{\alpha^n - \beta^n} = \frac{\alpha\left(1 - \left(\frac{\beta}{\alpha}\right)^{n+1}\right)}{1 - \left(\frac{\beta}{\alpha}\right)^n}. \qquad (3.15)$$

Using (3.13), we have

$$\left|\frac{\beta}{\alpha}\right| = \frac{|\alpha\beta|}{\alpha^2} = \frac{1}{\alpha^2} < 1,$$

and consequently the term $\left(\frac{\beta}{\alpha}\right)^n$ tends to zero as n tends to infinity. Then we see from (3.15) that the ratio F_{n+1}/F_n tends to the value α as n tends to infinity, which justifies our above assumption that F_{n+1}/F_n is close to α when n is large.

Problem 3.1.1 Deduce from (3.1) that $\gcd(F_{n+1}, F_n) = \gcd(F_n, F_{n-1})$, where $\gcd(a, b)$ denotes the greatest common divisor of a and b, and hence show that $\gcd(F_{n+1}, F_n) = 1$ for all $n \geq 1$.

Problem 3.1.2 Deduce directly from (3.7) that

$$\alpha + \beta = 1 \quad \text{and} \quad \alpha\beta = -1,$$

and also show that $\alpha - \beta = \sqrt{5}$.

Problem 3.1.3 Evaluate the sum of the first n Fibonacci numbers,

$$F_1 + F_2 + F_3 + \cdots + F_n,$$

for the first few values of n. Can you find a simpler way of expressing this sum?

3.2 Fibonacci and Lucas Numbers

There is another well-known sequence that satisfies the same recurrence relation as the Fibonacci sequence. This is the Lucas sequence, denoted by (L_n) and named after Edouard Lucas (1842–1891). The Lucas number L_n has the Binet form

$$L_n = \alpha^n + \beta^n, \qquad (3.16)$$

where α and β are given by (3.7). Using (3.13), we find that

$$L_1 = \alpha + \beta = 1,$$

$$L_2 = \alpha^2 + \beta^2 = (\alpha + \beta)^2 - 2\alpha\beta = 1 + 2 = 3, \tag{3.17}$$

and it follows from the recurrence relation that L_n is a positive integer for all $n \geq 1$. We will see in Section 4.3 how Lucas used the sequence (L_n) in a method he devised for testing whether a number of the form $2^p - 1$ is a prime, where p is itself a prime.

Example 3.2.1 From the Binet forms (3.12) and (3.16) for the Fibonacci and Lucas numbers, we have

$$F_n L_n = \left(\frac{\alpha^n - \beta^n}{\alpha - \beta} \right) (\alpha^n + \beta^n) = \frac{\alpha^{2n} - \beta^{2n}}{\alpha - \beta},$$

and thus obtain

$$F_n L_n = F_{2n}, \tag{3.18}$$

one of many identities that involve Fibonacci and Lucas numbers.   ■

n	1	2	3	4	5	6	7	8	9	10	11	12
F_n	1	1	2	3	5	8	13	21	34	55	89	144
L_n	1	3	4	7	11	18	29	47	76	123	199	322

TABLE 3.1. The first few members of the Fibonacci and Lucas sequences.

In Table 3.1, which displays the first few Lucas and Fibonacci numbers, we observe that each Lucas number is the sum of two Fibonacci numbers, the one to the right and the one to the left in the line above, which is saying that

$$L_n = F_{n+1} + F_{n-1}. \tag{3.19}$$

We can verify that this holds for $n = 2$ and $n = 3$. Let us *assume* that it is true for $n = m - 1$ and $n = m$, where $m \geq 3$. Then

$$L_{m+1} = L_m + L_{m-1} = (F_{m+1} + F_{m-1}) + (F_m + F_{m-2}),$$

and we can rearrange the terms involving the Fibonacci numbers to give

$$L_{m+1} = (F_{m+1} + F_m) + (F_{m-1} + F_{m-2}) = F_{m+2} + F_m.$$

We have used the recurrence relation for the Fibonacci numbers twice in the last step. Thus (3.19) holds for $n = m + 1$, and so by mathematical induction, it holds for all $n \geq 2$.

There is a wealth of identities involving Fibonacci or Lucas numbers. Some are more easily verified by induction, and some by algebraic manipulation, using the recurrence relation. In some cases it is simpler to begin by expressing each Fibonacci and Lucas number in its Binet form.

Example 3.2.2 Let us verify the identity

$$F_{n+1}F_{n-1} - F_n^2 = (-1)^n \tag{3.20}$$

by using the Binet form (3.12). Working with the numerator of the Binet form for the product $F_{n+1}F_{n-1}$, we obtain

$$(\alpha^{n+1} - \beta^{n+1})(\alpha^{n-1} - \beta^{n-1}) = \alpha^{2n} + \beta^{2n} - (\alpha\beta)^{n-1}(\alpha^2 + \beta^2),$$

and hence, using (3.13) and (3.17), we find that

$$(\alpha^{n+1} - \beta^{n+1})(\alpha^{n-1} - \beta^{n-1}) = \alpha^{2n} + \beta^{2n} + 3(-1)^n.$$

Similarly, we have from the numerator of the Binet form for F_n^2 that

$$(\alpha^n - \beta^n)^2 = \alpha^{2n} + \beta^{2n} - 2(-1)^n,$$

and on combining the last two results, we find that

$$(\alpha^{n+1} - \beta^{n+1})(\alpha^{n-1} - \beta^{n-1}) - (\alpha^n - \beta^n)^2 = 5(-1)^n.$$

Then, in view of (3.14), the identity (3.20) follows immediately on dividing the last equation throughout by $(\alpha - \beta)^2 = 5$. ∎

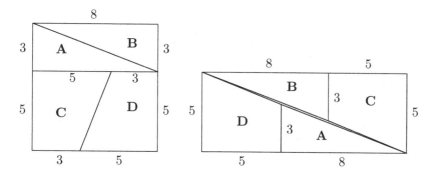

FIGURE 3.1. Geometrical interpretation of $F_{n+1}F_{n-1} - F_n^2 = (-1)^n$.

Consider a square with side F_n and a rectangle with sides F_{n+1} and F_{n-1}, illustrated in Figure 3.1 for the case $n = 6$. The identity (3.20) shows that for any value of $n \geq 2$, the areas of the square and rectangle differ by one square unit. Obviously, when n is large, one square unit is small compared with the two areas. If $n \geq 3$, we can write

$$F_n^2 = F_n(F_{n-1} + F_{n-2}) = F_n F_{n-1} + F_n F_{n-2}.$$

Geometrically, this is equivalent to dissecting the square of side F_n into two rectangles. See Figure 3.1, where both rectangles on the left are split into two equal pieces. The resulting four pieces, labeled A, B, C, and D, are rearranged to make the shape on the right side of Figure 3.1. This is an old party trick, cutting a square with side 8 into four pieces and rearranging them into a rectangular shape, as shown in Figure 3.1. If we don't do this too precisely, it can appear that a square whose area is 64 square units has been magically transformed into a rectangle whose area is 65 square units. Of course, there is a "hole" in the rectangle, consisting of a very narrow parallelogram whose area is 1 square unit.

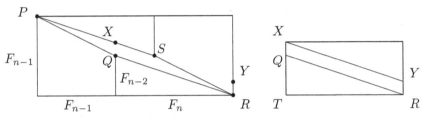

FIGURE 3.2. The area of $PQRS$ is 1, and $QX = 1/F_n$.

It is difficult to see the parallelogram in Figure 3.1 sufficiently well to analyze it further. Let us therefore look at Figure 3.2, which is labeled for the general case of n even, and whose dimensions correspond to the case $n = 4$. The parallelogram is labeled $PQRS$, and the point X is on PS, vertically above Q. The point Y is chosen vertically above R, so that $|RY| = |QX|$. It follows from this construction that the triangles PQX and SRY are congruent, and so the parallelogram $PQRS$ has the same area as the quadrilateral $QRYX$, which is also a parallelogram. From the diagram on the right side of Figure 3.2, which is a rectangle consisting of the parallelogram $QRYX$ plus two congruent right-angled triangles, we see that the area of the parallelogram $QRYX$ is the difference between the area $F_n \times |XT|$ and the area $F_n \times |QT|$. Thus we have

$$1 = \text{area of } QRYX = F_n \times |XT| - F_n \times |QT| = F_n \times |QX|,$$

which shows that $|QX|$, which is equal to the "height" of the parallelogram $PQRS$, is given by

$$|QX| = \frac{1}{F_n},$$

and we see that $|QX|$ decreases as n is increased. In particular, $|QX| = \frac{1}{8}$ in Figure 3.1.

Another identity with an obvious geometric interpretation is

$$\sum_{r=1}^{2n-1} F_r F_{r+1} = F_{2n}^2. \tag{3.21}$$

We will call a rectangle whose sides are consecutive Fibonacci numbers a *Fibonacci rectangle*. Figure 3.3, which illustrates the identity (3.21) for the case $n = 5$, shows a square of side 55 made up of the nine smallest Fibonacci rectangles, the smallest being the 1×1 square, and the largest the 34×55 rectangle. These rectangles may be put together to form the large square in different ways. You may wish to construct Fibonacci rectangles out of cardboard, and see how they may be assembled in other ways.

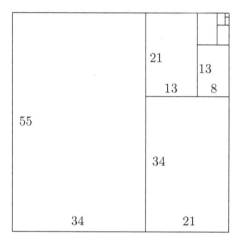

FIGURE 3.3. Geometrical illustration of the identity $\sum_{r=1}^{2n-1} F_{r+1}F_r = F_{2n}^2$.

If we choose an *even* number of Fibonacci rectangles, we can put them together to form a square with a 1×1 hole in it. This corresponds to the identity

$$\sum_{r=1}^{2n} F_r F_{r+1} = F_{2n+1}^2 - 1. \tag{3.22}$$

The geometrical justification of the identities (3.21) and (3.22) can be complemented by proofs using mathematical induction. For example, we see that (3.21) holds for $n = 1$, since $F_1 F_2 = F_2^2 = 1$. Note that when we replace n by $n + 1$, we add two terms to the sum on the left side of (3.21), and we can write

$$\sum_{r=1}^{2n+1} F_r F_{r+1} = F_{2n}F_{2n+1} + F_{2n+1}F_{2n+2} + \sum_{r=1}^{2n-1} F_r F_{r+1},$$

for any $n \geq 1$. Let us *assume* that (3.21) holds for some value of $n \geq 1$. Then the last equation shows us that

$$\sum_{r=1}^{2n+1} F_r F_{r+1} = F_{2n}F_{2n+1} + F_{2n+1}F_{2n+2} + F_{2n}^2. \tag{3.23}$$

Since the sum of the first and last terms on the right of (3.23) gives

$$F_{2n}F_{2n+1} + F_{2n}^2 = F_{2n}(F_{2n+1} + F_{2n}) = F_{2n}F_{2n+2},$$

it follows that the sum of all three terms on the right of (3.23) is

$$F_{2n}F_{2n+2} + F_{2n+1}F_{2n+2} = F_{2n+2}(F_{2n} + F_{2n+1}) = F_{2n+2}^2.$$

We therefore obtain from (3.23) that

$$\sum_{r=1}^{2n+1} F_r F_{r+1} = F_{2n+2}^2,$$

showing that the identity (3.21) also holds when n is replaced by $n + 1$. Thus, by mathematical induction, (3.21) holds for all $n \geq 1$. The identity (3.22) is proved similarly.

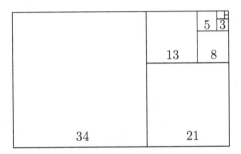

FIGURE 3.4. Geometrical illustration of the identity $\sum_{r=1}^{n} F_r^2 = F_n F_{n+1}$.

We have just seen how a Fibonacci square can be dissected into Fibonacci rectangles (with an additional 1×1 square left over for an odd-order Fibonacci square). There is a very pleasing converse result, that any Fibonacci rectangle can be dissected into Fibonacci squares. Specifically, we have

$$\sum_{r=1}^{n} F_r^2 = F_n F_{n+1}, \tag{3.24}$$

and this holds for all values of $n \geq 1$. Figure 3.4 illustrates this identity for the case $n = 9$, showing that the sum of the squares of the first 9 Fibonacci numbers is equal to $F_9 F_{10} = 34 \times 55$. Note that because $F_1 = F_2 = 1$, the two smallest squares in Figure 3.4 both have side 1. The identity (3.24) can be verified by induction. It holds for $n = 1$, since $F_1^2 = F_1 F_2 = 1$. We now write

$$\sum_{r=1}^{n+1} F_r^2 = F_{n+1}^2 + \sum_{r=1}^{n} F_r^2, \tag{3.25}$$

and assume that (3.24) holds for some $n \geq 1$. Then it follows from (3.24) that

$$\sum_{r=1}^{n+1} F_r^2 = F_{n+1}^2 + F_n F_{n+1} = F_{n+1}(F_{n+1} + F_n),$$

and on using the recurrence relation (3.1), we obtain

$$\sum_{r=1}^{n+1} F_r^2 = F_{n+1} F_{n+2}.$$

This shows that (3.24) also holds when n is replaced by $n + 1$, and so by mathematical induction, (3.24) holds for all $n \geq 1$.

Remark 3.2.1 Since I mentioned the word "parallelogram" in the geometrical analysis of the identity (3.20), this seems a good place in the text to make some remarks concerning the names of geometrical shapes. In English, many of these names closely resemble Greek or Latin words. Parallelogram is derived from Greek, and means "bounded by parallel lines." Triangle is derived from the Latin for "three corners," and the general four-sided figure is called a quadrilateral, after the Latin for "four sides." Pentagon is Greek for "five corners" or "five angles." We stick to Greek, using hexagon, heptagon, octagon, and decagon to denote shapes with six, seven, eight, and ten angles. The figure with nine angles is a nonagon, whose prefix comes from the Latin word *nonus*, meaning "ninth." Polygon comes from the Greek for "many angles," and trigonometry is derived from the Greek for the measurement of three-angled figures, better known to us as triangles. You may care to consult a dictionary to look up the word "square," whose derivation is less obvious. Of all the *regular* polygons (those that have all sides and angles equal), the square is the one we encounter most in everyday life. Perhaps this explains why in English and in the few other modern European languages whose word for "square" I happen to know, that word is not directly derived from Greek. If the naming of the square had been left to scholars, it might have been called the regular tetragon, which is too lengthy a title for such a familiar object.   ■

Example 3.2.3 Consider the equation

$$F_{m+n} = F_{m+1} F_n + F_m F_{n-1}. \tag{3.26}$$

We will justify (3.26) by induction. (For a combinatorial proof of (3.26) see Problem 5.1.5.) We see that (3.26) holds for $n = 2$ and all m, since this simply gives $F_{m+2} = F_{m+1} + F_m$. We also have

$$F_{m+3} = F_{m+2} + F_{m+1} = (F_{m+1} + F_m) + F_{m+1},$$

and hence

$$F_{m+3} = 2F_{m+1} + F_m,$$

showing that (3.26) holds for $n = 3$. Now let us assume that (3.26) holds for $2 \leq n \leq k$ and all m, for some $k \geq 3$. Thus, with $n = k - 1$ and $n = k$ we have

$$F_{m+k-1} = F_{m+1}F_{k-1} + F_mF_{k-2}$$

and

$$F_{m+k} = F_{m+1}F_k + F_mF_{k-1}.$$

On adding the last two equations "by columns," we immediately obtain

$$F_{m+k+1} = F_{m+1}F_{k+1} + F_mF_k.$$

Thus (3.26) holds for $n = k + 1$ and all m, and so by induction for $n \geq 2$ and all m. ■

The identity (3.26) helps us prove the following number-theoretical result concerning the Fibonacci numbers.

Theorem 3.2.1 For any integer $k \geq 1$, F_{nk} is divisible by F_n.

Proof. The theorem is obviously true when $k = 1$. Let us assume that it holds for some $k \geq 1$. Then, putting $m = kn$ in (3.26), we have

$$F_{(k+1)n} = F_{kn+1}F_n + F_{kn}F_{n-1}. \tag{3.27}$$

Since by our assumption $F_n \mid F_{kn}$, we see that F_n divides the right side of (3.27), and so divides $F_{(k+1)n}$. Thus, by induction, $F_n \mid F_{nk}$ for every integer $k \geq 1$. ■

Problem 3.2.1 Use the identity $x^2 - y^2 = (x - y)(x + y)$ to prove that

$$F_{n+1}^2 - F_n^2 = F_{n+2}F_{n-1},$$

for $n \geq 2$.

Problem 3.2.2 Use mathematical induction to verify that each of the following identities holds for all $n \geq 1$:

$$F_1 + F_2 + \cdots + F_n = F_{n+2} - 1,$$
$$F_1 + F_3 + \cdots + F_{2n-1} = F_{2n},$$
$$F_2 + F_4 + \cdots + F_{2n} = F_{2n+1} - 1.$$

Problem 3.2.3 Define

$$W_n = F_{n+1}F_{n-1} - F_n^2,$$

and verify that $W_2 = -W_3 = W_4 = 1$. Write

$$W_n + W_{n+1} = F_{n+1}F_{n-1} - F_n^2 + F_{n+2}F_n - F_{n+1}^2$$

and rearrange the four terms on the right of the latter equation to give

$$W_n + W_{n+1} = -F_{n+1}(F_{n+1} - F_{n-1}) + F_n(F_{n+2} - F_n).$$

Hence, by using (3.1), verify that

$$W_n + W_{n+1} = -F_{n+1}F_n + F_n F_{n+1} = 0,$$

and so show by induction that $F_{n+1}F_{n-1} - F_n^2 = (-1)^n$ for $n \geq 2$.

Problem 3.2.4 Using the relation $\alpha\beta = -1$, show that for $n > k \geq 0$,

$$(\alpha^{n+k} - \beta^{n+k})(\alpha^{n-k} - \beta^{n-k}) = \alpha^{2n} + \beta^{2n} + (-1)^{n-k+1}(\alpha^{2k} + \beta^{2k}),$$

so that with $k = 0$,

$$(\alpha^n - \beta^n)^2 = \alpha^{2n} + \beta^{2n} + 2(-1)^{n+1}.$$

Deduce that

$$F_{n+k}F_{n-k} - F_n^2 = (-1)^{n-k+1}F_k^2, \quad k > 0.$$

Problem 3.2.5 Show that

$$\alpha + \frac{1}{\alpha} = -\left(\beta + \frac{1}{\beta}\right) = \alpha - \beta,$$

and use the Binet form (3.12) to verify that

$$F_{n+1}^2 + F_n^2 = F_{2n+1}$$

for all integers $n \geq 1$.

Problem 3.2.6 Use the Binet form (3.16) to show that

$$L_{2n} = L_n^2 + 2(-1)^{n-1}.$$

Problem 3.2.7 Show by induction that

$$L_{n+1} + L_{n-1} = 5F_n.$$

Problem 3.2.8 Consider the sequence (m_k) given in (1.22), defined by

$$m_{k+1} = 6m_k - m_{k-1}, \quad k \geq 2, \quad \text{with} \quad m_1 = 1, \; m_2 = 6. \qquad (3.28)$$

Replace m_k by x^k and thus derive the characteristic equation

$$x^2 = 6x - 1.$$

Hence show that the general solution of the recurrence relation is

$$m_k = A(3 + 2\sqrt{2})^k + B(3 - 2\sqrt{2})^k,$$

and choose values of A and B such that $m_1 = 1$ and $m_2 = 6$, giving

$$m_k = \frac{1}{4\sqrt{2}}\left(\alpha^k - \alpha^{-k}\right), \quad \text{with} \quad \alpha = 3 + 2\sqrt{2},$$

as stated in (1.18).

Problem 3.2.9 By substituting $m_k = n_k + \frac{1}{2}$ in equation (3.28), verify that n_k is a solution of

$$n_{k+1} = 6n_k - n_{k-1} + 2, \quad k \geq 2.$$

Hence show that

$$n_k = A(3 + 2\sqrt{2})^k + B(3 - 2\sqrt{2})^k - \frac{1}{2},$$

and verify that the choice of $A = B = \frac{1}{4}$ gives a solution that matches the initial conditions $n_1 = 1$, $n_2 = 8$, in agreement with the solution given in (1.19).

Problem 3.2.10 Use the Binet forms (3.12) and (3.16) to show that

$$F_{(n+1)k} = L_k F_{nk} + (-1)^{k+1} F_{(n-1)k}.$$

Problem 3.2.11 The *harmonic mean* of two positive numbers a and b is defined as $2ab/(a+b)$. Show that the harmonic mean of F_n and L_n is F_{2n}/F_{n+1}.

Problem 3.2.12 Observe that the first three Fibonacci numbers are odd, odd, and even, respectively. Deduce from the recurrence relation that this pattern of odd, odd, and even is repeated for the whole sequence of Fibonacci numbers, so that F_{3n} is even for all n, and all other Fibonacci numbers are odd.

Problem 3.2.13 Follow the method used in Problem 3.2.12 to examine the remainders when Fibonacci numbers are divided by 3. Show that the pattern of the remainders is repeated in blocks of eight, and show that F_n is divisible by 3 if and only if n is divisible by 4.

Problem 3.2.14 Show that F_n is divisible by 5 if and only if n is divisible by 5.

Problem 3.2.15 Find the patterns that arise when we divide L_n by 2, 3, and 5, as we did for F_n in the previous three problems. In particular, show that no Lucas number is divisible by 5.

3.3   The Golden Section

If we divide (3.1) throughout by F_n, we obtain

$$\frac{F_{n+1}}{F_n} = 1 + \frac{F_{n-1}}{F_n}, \quad n \geq 2. \tag{3.29}$$

Let us write R_n to denote the ratio F_{n+1}/F_n. Then we see from (3.29) that

$$R_n = 1 + \frac{1}{R_{n-1}}, \quad n \geq 2. \tag{3.30}$$

On applying this repeatedly, we see that

$$R_n = \frac{F_{n+1}}{F_n} = [1, 1, \ldots, 1], \tag{3.31}$$

where 1 occurs n times in the continued fraction on the right of (3.31). We have already encountered the special case of (3.31) with $n = 8$, as equation (2.62), and we now see from (3.31) that F_{n+1}/F_n is the nth convergent of the infinite continued fraction

$$[1, 1, 1, \ldots] = \alpha = \frac{1 + \sqrt{5}}{2} \approx 1.618034. \tag{3.32}$$

Note that

$$\alpha = [1, 1, 1, \ldots] = [1, \alpha] = 1 + \frac{1}{\alpha},$$

so that, as we already know from our investigation of the solutions of (3.6),

$$\alpha^2 = \alpha + 1. \tag{3.33}$$

Since F_{n+1}/F_n is a convergent of an infinite continued fraction, we can apply the identity (2.46), which holds for consecutive convergents of any simple continued fraction. We thus obtain

$$F_{n+1}F_{n-1} - F_n^2 = (-1)^n,$$

as we obtained directly in Example 3.2.2 by using the Binet form. Table 3.2 displays some of the ratios F_{n+1}/F_n, rounded to three decimal places. The ratios corresponding to $n = 5$ and $n = 6$ are given exactly.

k	4	5	6	7	8	9	10
F_{n+1}/F_n	1.667	1.600	1.625	1.615	1.619	1.618	1.618

TABLE 3.2. Ratios of consecutive Fibonacci numbers.

The famous constant α defined by (3.32) is called the *golden section number* or the *golden ratio*. It is associated with a mathematical problem that goes back at least as far as the Pythagorean school of mathematics in the sixth century BC. The problem is to find the proportions of a rectangle such that if we remove from it a square whose side has the same length as the shorter side of the rectangle, the rectangle that remains has the same proportions as the original rectangle.

Since we are concerned only with proportions, let the shorter side be of length 1 and the longer side be of length x. Then, on equating the ratio

of the longer to the shorter sides in the large and small rectangles, we find
(see Figure 3.5) that x satisfies the equation

$$\frac{x}{1} = \frac{1}{x-1}.$$

On multiplying throughout by $x - 1$, we obtain

$$x(x - 1) = 1,$$

so that $x^2 = x + 1$. This verifies that the golden ratio is α, defined above
by (3.32).

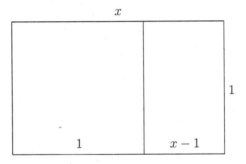

FIGURE 3.5. The golden section.

Golden rectangles were thought by the Pythagoreans to have an aesthetically pleasing shape, and it is said that the golden ratio occurs in the
dimensions of the Parthenon in Athens. The area of the whole rectangle in
Figure 3.5 is α, and that of the smaller rectangle is $\alpha - 1$. Thus the ratio
of their areas is

$$\frac{\alpha}{\alpha - 1} = \frac{\alpha^2}{\alpha^2 - \alpha} = \alpha^2,$$

since $\alpha^2 - \alpha = 1$. This is what we should expect, because the two rectangles
are similar. Beginning with Figure 3.5, we can remove a square from the
smaller golden rectangle, and repeat this process indefinitely. Obviously,
we can remove the square from either side of the rectangle at each stage.
Figure 3.6 gives one possible diagram obtained by carrying out this infinite
process. Since the areas of successive rectangles diminish by a factor of
$\alpha^2 \approx 2.618$, the areas of the rectangles tend to zero quite rapidly, and we
can see only a few of them. Figure 3.6, which contains an infinite number of
golden rectangles, is rather like Figure 3.4, which contains nine rectangles
whose sides are consecutive Fibonacci numbers.

The golden ratio also occurs in the star of Pythagoras, which is depicted
in Figure 3.7. This motif, which is also called a *pentagram*, was used as the

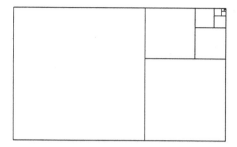

FIGURE 3.6. Repeated application of the golden section.

symbol of the Pythagoreans, the mathematicians who were associated with Pythagoras.

There are two obvious pentagons in Figure 3.7, the larger being the pentagon $ABCDE$, and the smaller being the interior polygon of which one side is FG. To emphasize the shape of the pentagram, the sides of the larger pentagon have not been drawn. The two pentagons are both *regular* pentagons, meaning that they are five-sided figures with all sides and angles equal. Any two of the points A, B, C, D, and E are either *adjacent*, like A and B or A and E, or not adjacent, like A and C or A and D. The larger pentagon consists of the straight lines connecting pairs of adjacent points, and the star is created by drawing straight lines connecting pairs of points that are not adjacent. Notice that we can draw the star by drawing one continuous line. For example, we can join A to C, C to E, E to B, B to D, and finally D to A.

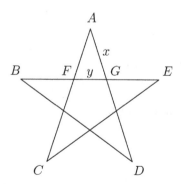

FIGURE 3.7. Star of Pythagoras.

Let us examine the angles that occur in Figure 3.7. There are only three different angles in the figure, one being an angle of the interior pentagon. Let us recall (see Problem 1.2.1) that the sum of the three angles of a triangle is 2π, and consider any *convex* polygon with n vertices. (A polygon is convex if a line joining any two points on its boundary lies entirely in the polygon.) Let us choose any point P in the interior of the polygon and join it to each of the n vertices. This creates n triangles, the sum of whose angles is $n\pi$. If we subtract the angles around the point P, whose sum is 2π, we see that the sum of the angles of the polygon is $(n-2)\pi$. If it is a *regular* polygon, then each angle is

$$\frac{(n-2)\pi}{n} = \left(1 - \frac{2}{n}\right)\pi. \tag{3.34}$$

As a check, we see that with $n = 3$ and $n = 4$, this correctly gives an angle $\frac{\pi}{3}$ for the equilateral triangle and $\frac{\pi}{2}$ for the angle in a square. With $n = 5$, we find that each angle of a regular pentagon is $\frac{3\pi}{5}$.

Knowing the size of the angle in a regular pentagon, we can determine the sizes of the other two angles that occur in Figure 3.7. An example of the larger of these is angle AFG which, together with the angle of the regular pentagon, adds up to 2π. Thus we have

$$\text{angle } AFG = \frac{2\pi}{5},$$

and there are ten angles of this size in Figure 3.7. Finally, an example of the smallest of the three angles in Figure 3.7 is angle FAG which, together with two copies of angle AFG, add up to two right angles. Thus

$$\text{angle } FAG = \frac{\pi}{5},$$

and there are five angles of this size in Figure 3.7, one at each point of the star. Triangle AFG, in which $|AF| = |AG|$, is called *isosceles*, as is any triangle that has two of its sides equal. We also see that $BCDG$ is a parallelogram, since its opposite sides are parallel, and thus $|BG| = |CD|$ and $|BC| = |GD|$. A further inspection of Figure 3.7 shows that all four sides of $BCDG$ are equal. Such a figure is called a *rhombus*. Its opposite angles are equal, and the square is the special case in which all four angles are equal. There are five copies of this rhombus in Figure 3.7. The lengths of all the line segments in Figure 3.7 can be expressed in terms of $|AG| = x$ and $|FG| = y$. Thus we have, for example,

$$|CD| = x + y \quad \text{and} \quad |AD| = 2x + y.$$

Now that we know all the angles in Figure 3.7, we can deduce so much more, since these angles determine everything but the scale of the figure. We see that triangles AFG and ACD are similar. Thus corresponding sides in these triangles have the same ratio, so that

$$\frac{x}{y} = \frac{|AG|}{|FG|} = \frac{|AD|}{|CD|} = \frac{2x+y}{x+y}. \tag{3.35}$$

We deduce from (3.35) that

$$\frac{x}{y} = \frac{2x+y}{x+y} = 1 + \frac{x}{x+y} = 1 + \frac{1}{1+y/x}. \tag{3.36}$$

On writing $t = x/y$, we have

$$t = 1 + \frac{1}{1+1/t} = 1 + \frac{t}{t+1} = \frac{2t+1}{t+1}.$$

If we multiply this last equation throughout by $t+1$, we obtain

$$t(t+1) = 2t+1,$$

so that $t^2 = t+1$, with $t = x/y$. This shows that $t = x/y = \alpha$, the golden ratio. We also have

$$\frac{|GD|}{|AG|} = \frac{x+y}{x} = 1 + \frac{y}{x} = 1 + \frac{1}{\alpha} = \alpha, \tag{3.37}$$

where the last equality is just (3.33) divided throughout by α. There are just four different lengths of line segments in the pentagram. In increasing order of size, these lengths are

$$y, \quad x, \quad x+y, \quad 2x+y. \tag{3.38}$$

In Figure 3.7 we have, for example, $y = |FG|$, $x = |AG|$, $x+y = |GD|$, and $2x+y = |AD|$. The pentagram contains five segments of length y, ten each of lengths x and $x+y$, and five of length $2x+y$. We see from (3.35) and (3.37) that

$$\frac{x}{y} = \frac{x+y}{x} = \frac{2x+y}{x+y} = \alpha. \tag{3.39}$$

Thus the ratios of consecutive members of the sequence in (3.38), the four different lengths of line segments in the pentagram, are *all* equal to the golden ratio! It is not surprising that the Pythagoreans were so enthusiastic about the pentagram.

Problem 3.3.1 Verify that the squares in Figure 3.6 have sides of length 1, $1/\alpha$, $1/\alpha^2$, $1/\alpha^3$, ..., and thus have areas 1, $1/\alpha^2$, $1/\alpha^4$, $1/\alpha^6$, ... Deduce from Figure 3.6 that

$$1 + \frac{1}{\alpha^2} + \frac{1}{\alpha^4} + \frac{1}{\alpha^6} + \cdots = \alpha.$$

Problem 3.3.2 In Figure 3.7 show that $|CD| = \alpha^2|FG|$, where α is the golden ratio, and deduce that the area of the outer pentagon, which connects the points of the star, is α^4 times the area of the inner pentagon. Show also that $\alpha^4 = 3\alpha + 2$.

Problem 3.3.3 Deduce from $\alpha^2 = \alpha + 1$ that

$$\alpha^3 = \alpha^2 + \alpha = 2\alpha + 1,$$

and use the relation $\alpha^n = \alpha^{n-1} + \alpha^{n-2}$, for $n > 2$, to show by induction that

$$\alpha^n = \alpha F_n + F_{n-1}, \quad n \geq 2.$$

Problem 3.3.4 Show by induction that

$$\frac{1}{\alpha^n} = (-1)^n (F_{n+1} - \alpha F_n),$$

for $n \geq 1$. Note that the quantity $1/\alpha^n$ tends to zero as n tends to infinity. What does the above equation tell you about the ratio F_{n+1}/F_n when n is large?

Problem 3.3.5 Deduce from the expression for $1/\alpha^n$ in Problem 3.3.4 that

$$\frac{F_{n+1}}{F_n} - \alpha$$

has the same sign as $(-1)^n$.

4
Prime Numbers

Mathematicians have tried in vain to this day to discover some order in the sequence of prime numbers, and we have reason to believe that it is a mystery into which the human mind will never penetrate.

Leonhard Euler (1707–1783)

4.1 Introduction

There are twenty-five prime numbers that are less than 100. These are displayed in Table 4.1.

2	3	5	7	11
13	17	19	23	29
31	37	41	43	47
53	59	61	67	71
73	79	83	89	97

TABLE 4.1. The twenty-five smallest primes.

A prime number (see Definition 1.1.1) is a positive integer greater than 1 that is divisible only by 1 and itself. I will now summarize the properties of the prime numbers that we have already met in this book.

In Section 1.4 it is stated that every positive integer n can be written uniquely in the form

$$n = p_1^{\alpha_1} p_2^{\alpha_2} p_3^{\alpha_3} \cdots p_N^{\alpha_N}$$

for some choice of N, where $p_1 = 2$, $p_2 = 3$, $p_3 = 5$, and so on, are the prime numbers. Each exponent α_1, α_2, $\alpha_3, \ldots \alpha_{N-1}$ is a nonnegative integer and the last exponent, α^N, is positive. Every odd prime is either of the form $4m+1$ or $4m+3$, and, as stated in Section 1.4, these are radically different in the following way:

1. Every prime number of the form $4m + 1$ can be expressed uniquely as the sum of two squares.

2. No prime number of the form $4m + 3$ can be expressed as the sum of two squares.

In Section 2.2 there is a proof of Fermat's little theorem, that for any prime number p and any positive integer a that is not divisible by p,

$$a^{p-1} \equiv 1 \pmod{p}.$$

In Section 2.4 we saw (Theorem 2.4.3) that if a prime p divides the product ab, where a and b are positive integers, then p divides at least one of a and b. In Section 2.5 I state that for any prime number p, $\sqrt{p}$ is irrational, and the proof I gave for $p = 2$ is easily adapted to the general case.

This completes the summary of results concerning prime numbers that we have already met. We continue with two very important results that have been known to mathematicians for two millennia. The first is a proof that there is an infinite number of prime numbers. The second is a systematic method for *finding* them.

Theorem 4.1.1 The number of primes is infinite.

Proof. This most elegant proof was given by Euclid circa 300 BC. We begin by assuming that the number of primes is finite. Let us denote them by $p_1, p_2, \ldots, p_n$, so that $p_1 = 2$, $p_2 = 3$, $p_3 = 5$, and so on. Now consider the number

$$q = p_1 p_2 \cdots p_n + 1, \tag{4.1}$$

where we have multiplied together all the primes, and added 1. This integer q cannot be a prime, since it is larger than all n primes. Thus q must have a prime factor, say p, which must be one of the primes $p_1, p_2, \ldots, p_n$. But it is clear from (4.1) that none of these primes divides q. This contradicts the initial assumption that the number of primes is finite. ∎

Eratosthenes (circa 230 BC) is remembered for outstanding work in two quite distinct fields. For in addition to his mathematical contributions, he obtained good estimates both of the diameter of the earth and of the angle of tilt of the earth's axis. In mathematics, his most enduring achievement is the so-called *sieve of Eratosthenes*, a systematic method for finding all prime numbers less than a given positive integer n. Eratosthenes wrote down the numbers 1, 2, 3, and so on, and identified 2, the first number after 1, as the first prime. He then crossed out every second number after 2, so eliminating 4, 6, 8, and so on. None of these can be a prime, since they are all divisible by 2. He then has before him the equivalent of Table 4.2, where the number 2 is underlined to signify that it has been identified as a prime. (I have placed a row of dots in the third line of the table to indicate that the table can be extended as far as we wish.)

	1	2	3	4	5	6	7	8	9
10	11	12	13	14	15	16	17	18	19
.	.	.	.	.	.	.	.	.	.

TABLE 4.2. The sieve of Eratosthenes: cross out multiples of 2.

Eratosthenes then identified 3, the first number after 2 that is not crossed out, as the second prime. He crossed out every third number after 3. Since these are all divisible by 3, none of these is a prime. (Of course, those multiples of 3 that are also multiples of 2 are already crossed out.) Eratosthenes continued in this way. At each stage he identified the next prime, say p, and then crossed out every multiple of p. Table 4.3 shows the outcome after crossing out multiples of 2, 3, 5, and 7.

	1	2	3	4	5	6	7	8	9
10	11	12	13	14	15	16	17	18	19
20	21	22	23	24	25	26	27	28	29
30	31	32	33	34	35	36	37	38	39
40	41	42	43	44	45	46	47	48	49
50	51	52	53	54	55	56	57	58	59
60	61	62	63	64	65	66	67	68	69
70	71	72	73	74	75	76	77	78	79
80	81	82	83	84	85	86	87	88	89
90	91	92	93	94	95	96	97	98	99
.	.	.	.	.	.	.	.	.	.

TABLE 4.3. The sieve of Eratosthenes: cross out multiples of 2, 3, 5, and 7.

An inspection of Table 4.3 tells us that 11, the first number that is not underlined and is not crossed out, is the next prime after 7. However, we can say much more than this. Let us suppose that Table 4.3 is extended as far as we wish. Then every number that is not crossed out and is *not* a prime must be the product of factors that are not less than 11, and there must be at least two factors. Thus all uncrossed numbers less than 11×11 must be primes, and we can infer the following general result.

Theorem 4.1.2 If we use the sieve of Er-
atosthenes, we have crossed out multiples of the first $n - 1$ primes, p_1, $p_2, \ldots, p_{n-1}$. Then the first number after p_{n-1} that is not crossed out is the nth prime, p_n, and *every* number in the table less than p_n^2 that is not crossed out is a prime. ∎

Note that we can find all primes up to some value N by using the sieve of Eratosthenes to cross out multiples of all primes not greater than $\sqrt{N}$. This is what makes the process so powerful. It is rather tedious to implement by hand, because it is so easy to make a mistake. Imagine crossing out multiples of 43, for example. However, the sieve of Eratosthenes is a wonderful algorithm and was there, ready for serious action, when the automatic digital computer appeared in the middle of the twentieth century.

Example 4.1.1 If we use the sieve of Eratosthenes to cross out multiples of the first twenty-five primes $2, 3, 5, \ldots, 97$, we obtain the 1229 primes that are less than less than 10,000, since $p_{26}^2 = 101^2 > 10,000$. To find all the 78,498 primes that are less than one million, we use 168 applications of the sieve, since $p_{168} = 997$ and $p_{169}^2 = 1009^2 > 1,000,000$. ∎

Problem 4.1.1 Let $q = p_1 p_2 \cdots p_n + 1$, as defined in Theorem 4.1.1, where $p_1, p_2, \ldots, p_n$ denote the first n primes. Evaluate q for the first few values of n and find the smallest value of n for which q is not a prime.

Problem 4.1.2 Assume that there is a finite number of primes of the form $4m + 3$. Denote them by $q_1, q_2, \ldots, q_k$, where $q_1 = 3$ and $q_2 = 7$, and write

$$q = 4q_1 q_2 \cdots q_k - 1.$$

Verify that q is of the form $4m + 3$. Then, in the spirit of Theorem 4.1.1, deduce that q cannot be prime and so must have a prime factor, which must be odd. Deduce from Problem 2.2.2 that the prime factors of q cannot all be of the form $4m + 1$, and thus q must have at least one prime factor of the form $4m + 3$. Show that this gives a contradiction.

4.2   The Distribution of Primes

A detailed inspection of any portion of a table of prime numbers reveals no obvious pattern. There is no known simple expression that we can evaluate to give us a practical way of generating all the prime numbers. Mathematicians would be very pleased if they could even find an increasing function, say $F(n)$, such that $F(n)$ is a prime for all positive integer values of n. No such expression is known. Some simple functions take prime values surprisingly often. For example, the quadratic polynomial $n^2 - n + 41$ has prime number values for 40 consecutive values of n, beginning with $n = 1$, but then has the value 41^2 for $n = 41$.

n	10^2	10^3	10^4	10^5	10^6	10^7
$\pi(n)$	25	168	1229	9592	78,498	664,579

TABLE 4.4. Values of $\pi(n)$, the number of primes not greater than n.

A longstanding goal of mathematicians, which was effectively attained by the end of the nineteenth century, was to determine how many primes there are in the first n positive integers. We intuitively expect, from the way the sieve of Eratosthenes operates, that the primes thin out as n is increased. This hypothesis is supported by numerical evidence. Let $\pi(n)$ denote the number of primes that are not greater than n. For example, $\pi(10) = 4$, and some further values of $\pi(n)$ are given in Table 4.4. The largest prime less than 10^7 is 9,999,991.

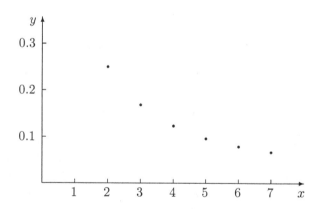

FIGURE 4.1. The proportion of primes in the first 10^x numbers.

We see immediately that the ratio $\pi(n)/n$ decreases slowly for the values of n given in Table 4.4, from 0.25 for $n = 10^2$ to a little more than 0.066 for $n = 10^7$. Let us write

$$y = \frac{\pi(10^x)}{10^x}.$$

Figure 4.1 shows the values of y for integer values of x between 2 and 7, as given in Table 4.4. This pictorial evidence suggests that when we look at sufficiently large blocks of primes, there is an orderliness that is not visible when we look at a table of primes in detail.

There is a gap of 1 between 2 and the next prime, 3. Apart from 2, all primes are odd, and thus there must be a gap of at least 2 between successive primes from 3 onwards. We observe that there is a gap of 2 between 3 and 5, 5 and 7, 11 and 13. The pairs $(3, 5)$, $(5, 7)$, and $(11, 13)$ are called *prime pairs* or *twin primes*. Let $\pi_2(n)$ denote the number of pairs $(p, p+2)$ where p and $p+2$ are both primes, with $p+2 \le n$. There are two such pairs below 10, namely $(3, 5)$ and $(5, 7)$, and so $\pi_2(10) = 2$. Numerical evidence suggests that there is an infinite number of prime pairs, but no proof is known at the present time. Mathematicians would dearly like to settle the question of the infinitude of prime pairs. I should add that when we look at sufficiently large blocks of prime pairs, there is an orderliness in the behavior of $\pi_2(n)/n$ for large n that is just as convincing as that found in the behavior of $\pi(n)/n$ for large n. However, at present, although the regularity of the behavior of $\pi(n)/n$ has been justified by the prime number theorem stated below, there is no comparable theorem known so far about $\pi_2(n)/n$, although there is a strong conjecture about it that I also quote below. Some values of the function $\pi_2(n)$ are given in Table 4.5. The largest prime pair less than 10^7 is $(p, p+2)$, with $p = 9{,}999{,}971$.

n	10^2	10^3	10^4	10^5	10^6	10^7
$\pi_2(n)$	8	35	205	1224	8169	58,980

TABLE 4.5. Values of $\pi_2(n)$, the number of prime pairs not greater than n.

We have seen that apart from the gap between the first two primes, 2 and 3, the next smallest gap of 2 (the property that defines prime pairs) appears to occur with a frequency that diminishes slowly as we search through the list of primes. What about larger gaps? Consider the sequence of $n-1$ numbers $n!+2$, $n!+3$, ..., $n!+n$. They are all composite, because the first number is divisible by 2, the second by 3, and so on, and the last number, $n! + n$, is divisible by n. This simple observation yields the remarkable result that there are gaps between consecutive primes as large as we please. Somewhere, in the unending list of primes, there occurs for the first time a gap between consecutive primes that exceeds one million.

C. F. Gauss made substantial discoveries in all the areas of mathematics that were known in his time, as well as making serious contributions to astronomy and physics. Within mathematics he had a particular love of number theory, as is clear from his well-known remark, "Mathematics is the queen of the sciences, and the theory of numbers is the queen of mathematics." From his study of tables of primes, Gauss formed a conjecture about the approximate value of $\pi(n)$ for large n. He was very skilful in carrying out calculations, and these sometimes helped him form conjectures, which he usually succeeded in proving. Gauss's conjecture about the distribution of the primes is expressed in terms of the *logarithmic* function, which we write as $\log x$ and define as the area above the t-axis, below the *hyperbola* $y = 1/t$, and between the ordinates $t = 1$ and $t = x \geq 1$. (See Figure 4.2.)

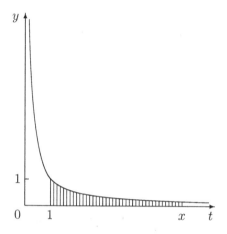

FIGURE 4.2. The area under the hyperbola $y = 1/t$ between 1 and $x \geq 1$ is the logarithm of x.

Those who know about *integration* will be familiar with the expression of this area as the integral

$$\log x = \int_1^x \frac{dt}{t}, \tag{4.2}$$

and be able to deduce that

$$\int_1^{x_1 x_2} \frac{dt}{t} = \int_1^{x_1} \frac{dt}{t} + \int_1^{x_2} \frac{dt}{t}, \tag{4.3}$$

for $x_1, x_2 \geq 1$. It then follows from (4.2) and (4.3) that

$$\log x_1 x_2 = \log x_1 + \log x_2, \tag{4.4}$$

which is the characterizing property of the logarithm. I state without proof that the logarithm is the *inverse* function of the *exponential* function. This means that if $y = \log x$, where $x > 0$, then

$$x = e^y = 1 + \frac{y}{1!} + \frac{y^2}{2!} + \frac{y^3}{3!} + \cdots . \tag{4.5}$$

Although the series for e^y has been defined above only for real values of y, it also has a meaning for all complex values. The above series converges (and so can be used to compute e^y to any required accuracy) for *all* complex values of y. The exponential function has the well-known property that

$$e^{y_1} e^{y_2} = e^{y_1 + y_2} \tag{4.6}$$

for all complex values of y_1 and y_2.

We also require the concept of *asymptotic* equality. If the ratio of $f(x)$ and $g(x)$ tends to 1 as x tends to infinity, we say that $f(x)/g(x)$ has the *limit* 1, and write

$$\lim_{x \to \infty} \frac{f(x)}{g(x)} = 1. \tag{4.7}$$

We then say that $f(x)$ and $g(x)$ are *asymptotically equal* for large x, and write this as

$$f(x) \sim g(x). \tag{4.8}$$

Let us now turn to the approximate value of $\pi(n)$ for large n. Gauss conjectured that

$$\pi(x) \sim \frac{x}{\log x}. \tag{4.9}$$

This conjecture was later shown to be correct, and is called the *prime number theorem*. Equivalently, Gauss conjectured that

$$\pi(x) \sim \mathrm{Li}(x) = \int_2^x \frac{dt}{\log t}, \tag{4.10}$$

where the function $\mathrm{Li}(x)$ is called the *logarithmic integral*. Thus $\mathrm{Li}(x)$ is the area above the t-axis, below the curve $y = 1/\log t$, and between the ordinates $t = 2$ and $t = x \geq 2$. I say "equivalently" in comparing (4.9) and (4.10) because it can be shown that

$$\mathrm{Li}(x) \sim \frac{x}{\log x}. \tag{4.11}$$

n	10^3	10^6	10^9
$\pi(n)$	168	78,498	50,847,478
$n/\log n$	0.14476×10^3	0.72382×10^5	0.48255×10^8
$\mathrm{Li}(n)$	0.17656×10^3	0.78627×10^5	0.50849×10^8

TABLE 4.6. Distribution of the prime numbers.

Table 4.6 displays values of $\pi(n)$, $n/\log n$, and $\mathrm{Li}(n)$ for $n = 10^3$, 10^6, and 10^9. The values of $\pi(n)$ are exact, and the other values are given to five significant digits. We note that, at least in this table,

$$|\pi(n) - \mathrm{Li}(n)| < |\pi(n) - n/\log n|,$$

and it is quite striking that for $n = 10^9$, $\mathrm{Li}(n)$ and $\pi(n)$ agree to four significant digits.

Mathematics can defeat even the mightiest, and Gauss never *proved* the prime number theorem. However, P. L. Chebyshev (1821–1894) made considerable progress towards this goal. In 1896, not long after Chebyshev's death, J. Hadamard (1865–1963) and C. J. de la Vallée Poussin (1866–1962) both found proofs independently. These two long-lived mathematicians are indeed a very special "prime pair," and if Euler could have seen one of their proofs I wonder whether he would have retracted the statement that is quoted at the head of this chapter.

Although the prime number theorem is concerned with integers and real numbers, the proofs published in 1896 used complex variable theory. "Elementary" proofs of the prime number theorem that do not use complex numbers were published independently in 1949 by Paul Erdős (1913–1996) and Atle Selberg (born 1917). In his delightful book [13] about Paul Erdős, Paul Hoffman alludes in rhyme to a simpler theorem proved by Chebyshev, which settled an earlier conjecture of Joseph Bertrand (1822–1900):

> Chebyshev said it, and I say it again
> There is always a prime between n and $2n$.

Hardy and Wright [12] state the following conjecture, which gives an asymptotic estimate of $\pi_2(n)$, the number of prime pairs not exceeding n:

$$\pi_2(n) \sim \frac{Cn}{(\log n)^2}, \tag{4.12}$$

where C is the constant defined by

$$C = 2 \prod_{p \geq 3} \left(1 - \frac{1}{(p-1)^2} \right). \tag{4.13}$$

The symbol Π on the right of (4.13) is the uppercase Greek letter *pi*, and it denotes the *product* taken over all prime numbers p from 3 onwards. Of course, a *proof* that (4.12) holds would immediately tell us that there is an infinite number of prime pairs.

There are many conjectures about prime numbers that are easily stated, but which have so far remained unproved, including the following.

- Every even number greater than 4 can be expressed as the sum of two odd primes. This is Goldbach's conjecture, named after C. Goldbach (1690–1764).

- There is at least one prime between any two consecutive squares.

- There is an infinite number of primes of the form $n^2 + 1$.

- There is an infinite number of primes of the form $n! + 1$.

- There is an infinite number of primes of the form $p_1 p_2 \cdots p_n + 1$.

The best-known of the above conjectures is that of Goldbach, which is also the one whose resolution has been the most sought.

Having mentioned so many conjectures that are unproved at the time I am writing this, I am pleased to mention a conjecture that was settled in 2004. The prime numbers 3, 5, and 7 are evenly spaced. We say that they form an arithmetical progression with common difference 2. The next such sequence of three prime numbers in arithmetical progression is 31, 37, and 43, with common difference 6. Johannes van der Corput showed in 1939 that there are infinitely many sequences of three prime numbers in arithmetical progression. Ben Green and Terence Tao sought to prove that there are infinitely many sequences of four prime numbers in arithmetical progression. However, in 2004 they succeeded in proving very much more than that: they proved that given *any* positive integer n, there are infinitely many sequences of n prime numbers in arithmetical progression. For example, there are ten prime numbers in arithmetical progression beginning with 199, with common difference 210, so that the tenth number is 2089, and Green and Tao's result tells us that that are infinitely many sequences of 10 prime numbers in arithmetical progression.

Problem 4.2.1 Evaluate the quadratic polynomial $p(n) = n^2 - n + 41$ for the values $n = 1, 2, \ldots, 41$. Verify that $p(n)$ has prime values for $1 \le n \le 40$, and has the value 41^2 when $n = 41$.

Problem 4.2.2 The polynomial $q(n) = n^2 - 79n + 1601$ is correctly described in several books as having prime values for $0 \le n \le 79$. Show that $q(79 - n) = q(n)$ for all n, and thus only 40 different primes are obtained for $0 \le n \le 79$. Verify that $q(n) = p(40 - n)$, where $p(n)$ is the quadratic polynomial defined in Problem 4.2.1.

Problem 4.2.3 Prove that the numbers p, $p + 2$ and $p + 4$ are all prime only when $p = 3$.

Problem 4.2.4 In the text we state that if $y = \log x$, for $x > 0$, then $x = e^y$, where e^y is defined by (4.5). Deduce that $x = e^{\log x}$, for $x > 0$.

Problem 4.2.5 Argue that because the function $y = 1/\log t$ is decreasing, the area under the curve $y = 1/\log t$ from $t = 2$ to $t = n$ is greater than the area of a rectangle of width $n - 2$ and height $1/\log n$, so that

$$\text{Li}(x) = \int_2^n \frac{dt}{\log t} > \frac{n - 2}{\log n} \sim \frac{n}{\log n}.$$

Note that this is not sufficient to justify (4.11).

Problem 4.2.6 Check how well the two sides of (4.12) agree for the values of $\pi_2(n)$ given in Table 4.5, given that the constant C is approximately 1.32.

4.3   Large Primes

We begin by deriving an identity that we will use later in this section. Consider the *geometric series*

$$S = 1 + x + x^2 + \cdots + x^{n-1}, \tag{4.14}$$

where n is a positive integer. On multiplying (4.14) throughout by x, we find that

$$xS = x + x^2 + x^3 + \cdots + x^n. \tag{4.15}$$

If we subtract (4.15) from (4.14), all but two of the terms cancel to give

$$(1 - x)S = 1 - x^n, \tag{4.16}$$

so that if $x \neq 1$, the sum of the geometric series can be expressed as

$$S = \frac{1 - x^n}{1 - x}, \tag{4.17}$$

or in the form

$$S = \frac{x^n - 1}{x - 1}. \tag{4.18}$$

The latter form is preferred when $x > 1$, for then S is expressed as a fraction in which both numerator and denominator are positive. If $-1 < x < 1$, we can allow n to tend to infinity in (4.14). Then x^n tends to zero, and we obtain from (4.17) that

$$1 + x + x^2 + \cdots = \frac{1}{1 - x}, \tag{4.19}$$

where the dots on the left of (4.19) indicate that we have an infinite series. If $|x| \geq 1$, the infinite series has no sum and (4.19) does not hold.

Let us return to (4.16), which can be written as

$$1 - x^n = (1 - x)\left(1 + x + x^2 + \cdots + x^{n-1}\right). \tag{4.20}$$

If we replace x by b/a in (4.20), and multiply throughout by a^n, we obtain

$$a^n - b^n = (a - b)\left(a^{n-1} + a^{n-2}b + \cdots + ab^{n-2} + b^{n-1}\right), \tag{4.21}$$

and this holds for all $n \geq 2$. Although this identity can be verified directly by multiplying the two factors on the right of (4.21), I wanted to show how it is connected to the sum of a geometric series.

Pierre de Fermat defined the number

$$f_n = 2^{2^n} + 1, \tag{4.22}$$

which we now call a *Fermat number*. He conjectured that f_n is a prime for all positive integers n. In fact, this conjecture is *false*. For although

$$f_1 = 5, \quad f_2 = 17, \quad f_3 = 257, \quad f_4 = 65{,}537$$

are indeed primes, Leonhard Euler showed in 1732 that

$$f_5 = 4{,}294{,}967{,}297 = 641 \times 6{,}700{,}417,$$

and thus f_5 is *not* prime.

The Fermat numbers grow very rapidly. Since

$$f_{n+1} \approx 2^{2^{n+1}} = 2^{2^n} \times 2^{2^n} \approx f_n^2,$$

f_{n+1} has approximately twice as many digits as f_n. For example, f_5 and f_6 have, respectively, ten and twenty decimal digits. Several further Fermat numbers have been factorized to date, and no other prime Fermat number has been found to add to the four listed above. Although the Fermat numbers are not all primes, they do at least share one very important property with the prime numbers, as given in the following theorem.

Theorem 4.3.1 No two Fermat numbers have a common divisor greater than 1.

Proof. We begin with (4.21), putting $b = -1$ and $n = 2^k$ with $k \geq 1$, to give

$$a^{2^k} - 1 = (a + 1)\left(a^{2^k-1} - a^{2^k-2} + \cdots + a - 1\right). \qquad (4.23)$$

The terms in the second factor on the right of (4.23) are alternately positive and negative. We note from (4.23) that if a and k are positive integers, $a+1$ is a divisor of $a^{2^k} - 1$. Now let us put $a = 2^{2^m}$, so that

$$a^{2^k} = \left(2^{2^m}\right)^{2^k} = 2^{2^m \cdot 2^k} = 2^{2^{m+k}} = f_{m+k} - 1.$$

Then we have

$$a^{2^k} - 1 = f_{m+k} - 2 \quad \text{and} \quad a + 1 = f_m,$$

and we see from (4.23) that

$$f_m \mid f_{m+k} - 2, \qquad (4.24)$$

for all positive integers m and k. We deduce from (4.24) that any number that divides both f_m and f_{m+k} must also divide 2. Since every Fermat number is odd, any common divisor of f_m and f_{m+k} must be odd, and thus their greatest common divisor must be 1. ∎

Definition 4.3.1 Recall that we write $\gcd(a, b)$ to denote the greatest common divisor of a and b. If $\gcd(a, b) = 1$, we say that a and b are *coprime*. The numbers a_1, a_2, ... are said to be coprime if $\gcd(a_j, a_k) = 1$ for all j and k. Note that when there are more than two numbers, saying that they are coprime is a much stronger statement than saying that their greatest common divisor is 1.   ■

Remark 4.3.1 The Fermat numbers and the prime numbers are both infinite sets of numbers that are coprime. In contrast, although we saw from Problem 3.1.1 that any two *consecutive* Fibonacci numbers are coprime, Theorem 3.2.1 shows that this is certainly not true of *all* pairs of Fibonacci numbers, and so the Fibonacci numbers are not coprime.   ■

Marin Mersenne (1588–1648), a French monk who corresponded with many of the leading scientists of his time, including Pierre de Fermat and Galileo Galilei (1564–1642), studied numbers of the form $2^n - 1$. Although many others had considered these numbers before him, they are called *Mersenne numbers*, and we will write

$$M_n = 2^n - 1. \tag{4.25}$$

Mersenne knew that M_n cannot be a prime if n is composite. For if we put $a = 2^k$ and $b = 1$ in (4.21), and also replace n by m, we see that

$$a^m = \left(2^k\right)^m = 2^{km},$$

and then (4.21) becomes

$$2^{km} - 1 = \left(2^k - 1\right)\left(a^{m-1} + a^{m-2}b + \cdots + ab^{m-2} + b^{m-1}\right). \tag{4.26}$$

I have not substituted $a = 2^{km}$ and $b = 1$ in the second factor on the right of (4.26) because it is not necessary to do so. For it is clear from (4.26) that $2^k - 1$ is a factor of $2^{km} - 1$. Thus if we are seeking prime numbers of the form $2^n - 1$, we must restrict our search to prime values of the exponent n. The first few Mersenne primes are displayed in Table 4.7.

n	2	3	5	7	13	17	19	31
M_n	3	7	31	127	8191	131,071	524,287	2,147,483,647

TABLE 4.7. The first few Mersenne primes.

Mersenne stated that M_n is prime for

$$n = 2, 3, 5, 7, 13, 17, 19, 31, 67, 127, 257$$

and for no other values of $n \leq 257$. It is now known that there are five errors in Mersenne's assertion. Two on his list, M_{67} and M_{257}, are *not* primes, and he omitted M_{61}, M_{89}, and M_{107}, which *are* primes. Mersenne was at least correct in implying that primes of this type are not plentiful. The smallest value of the prime p for which $2^p - 1$ is not prime is $p = 11$, and it is easily verified that

$$2^{11} - 1 = 2047 = 23 \times 89.$$

The eighth Mersenne prime, M_{31}, was found by Euler in 1772. Only two further Mersenne primes were found before 1900, namely M_{127} in 1876 and M_{61} in 1883. Note that M_{127} was discovered by Edouard Lucas, not only before the discovery of M_{61}, but also before the discoveries of M_{89} and M_{107}, which did not happen until 1911 and 1914, respectively, bringing the total of Mersenne numbers known by 1914 to twelve. In more recent years, the steady growth in computing power has made it possible to find increasingly larger Mersenne primes. It is still not known whether there is an infinite number of these. If anyone asks, "What practical use is there in finding such extremely large prime numbers?" a reasonable one-word answer would be "None!" For although the product of two large primes has been made the basis of the encryption of confidential information, the size of such "large" primes has been many tens of decimal digits, compared with the many millions of decimal digits that measure the size of the largest known Mersenne primes.

At the time of writing, 41 Mersenne primes have been found. The latest Mersenne prime, whose discovery was announced in May 2004, is M_n, with $n = 24{,}036{,}583$. This is the largest known prime of any kind. The previous record holder, found in 2003, is M_n with $n = 20{,}996{,}011$. Since

$$2^{10} = 1024 \approx 10^3$$

we can see that M_n with $n = 24{,}036{,}583$ has more than seven million decimal digits. If we were to type out this number on a ribbon of paper, in decimal form with three digits per centimeter, the ribbon would stretch for more than 20 kilometers. The seven most recently discovered Mersenne primes were all obtained by the Great Internet Mersenne Prime Search (GIMPS), involving the cooperation of more than 100,000 individuals.

The reason that so much of the effort in the search for large primes has concentrated on Mersenne numbers is due to the discovery of a very clever and highly efficient procedure by Edouard Lucas, which he used to establish that M_{127} is a prime in 1876. This makes the testing of a Mersenne number for primality substantially easier than the testing of a typical number of a similar size. In its present slightly revised form, this procedure is called the Lucas–Lehmer test. It is expressed in the following theorem.

Theorem 4.3.2 (Lucas–Lehmer test) Let (r_k) denote the sequence defined recursively by

$$r_{k+1} = r_k^2 - 2, \quad \text{for } k \geq 1, \quad \text{with } r_1 = 4. \tag{4.27}$$

Then M_p is a prime if and only if

$$r_{p-1} \equiv 0 \,(\text{mod } M_p),$$

that is, r_{p-1} is divisible by M_p.   ■

The proof of Theorem 4.3.2 is omitted, because it is too advanced for this book. Lucas used the above sequence (r_k) for primes of the form $4m + 1$ and used a similar sequence (r_k) with initial value $r_1 = 3$ for primes of the form $4m + 3$. D. H. Lehmer (1905–1991) showed that the sequence beginning with $r_1 = 4$ can be used for both sets of primes, and this is why the names of both Lucas and Lehmer are associated with Theorem 4.3.2.

The Standards Western Automatic Computer (SWAC), built for the U.S. National Bureau of Standards, came into service in 1950 at the University of California, Los Angeles. With the involvement of D. H. Lehmer, it was the first automatic digital computer to be used successfully in the hunt for Mersenne primes. In 1952 no fewer than five further Mersenne primes were found, namely M_{521}, M_{607}, M_{1279}, M_{2203}, and M_{2281}, bringing the total number of known Mersenne primes to eighteen. This statistic of five primes in one year stands out in the historical record of the Mersenne primes to date, an achievement made possible by bringing together greater computing power and the efficiency of the Lucas–Lehmer test.

On the subject of computing I should point out that the *binary* representation of M_n could hardly be simpler, since it consists of the digit 1 repeated n times. (See (2.10).) For example, $M_{13} = 8191 = (1,111,111,111,111)_2$.

Example 4.3.1 When we use the Lucas–Lehmer test we compute the smallest residue of each r_k modulo M_p, rather than r_k itself. It is convenient to continue to use r_k to denote this residue, which is therefore less than M_p. At each stage we have to compute $r_{k+1} = r_k^2 - 2$, which will be less than M_p^2, and find its smallest residue modulo M_p. Let us apply the Lucas–Lehmer test to $M_5 = 31$. We find that

$$r_1 = 4, \quad r_2 = 14, \quad r_3 = 194 \equiv 8 \,(\text{mod } 31), \quad r_4 = 62 \equiv 0 \,(\text{mod } 31),$$

showing that M_5 is a prime. Of course, it is trivial to check *directly* that $M_5 = 31$ is a prime number. To give a more realistic impression of the Lucas–Lehmer test, let us also check that $M_{19} = 524{,}287$ is a prime. The first 18 values of the sequence (r_k) are given in Table 4.8.

$r_1 = 4$	$r_2 = 14$	$r_3 = 194$	$r_4 = 37{,}634$	218,767	510,066
$r_7 = 386{,}344$	323,156	218,526	504,140	103,469	417,706
$r_{13} = 307{,}417$	382,989	275,842	85,226	523,263	$r_{18} = 0$

TABLE 4.8. The residues of the sequence (r_k) obtained when the Lucas–Lehmer test is applied to the Mersenne number M_{19}.

As we write down the values r_1, r_2, and so on, then from $r_5 = 218{,}767$ onwards, the numbers in the sequence seem to pop up anywhere in the possible range from 0 to $M_{19} - 1 = 524{,}286$. Yet, magically, r_{18} has the value 0, showing that M_{19} is indeed a Mersenne prime.   ∎

The above numerical example helps us to appreciate more fully the magnitude of what was achieved in the precomputer era, notably by Euler in checking that the Mersenne number $M_{31} = 2{,}147{,}483{,}647$ is a prime and by Lucas in checking that the 39-digit Mersenne number

$$M_{127} = 170{,}141{,}183{,}460{,}469{,}231{,}731{,}687{,}303{,}715{,}884{,}105{,}727$$

is a prime. Although Lucas had the advantage of using his test, described in Theorem 4.3.2, he still had to carry out a rather onerous calculation by hand. At each stage Lucas had to take a number r_k of up to 39 decimal digits, square it, subtract 2 to give r_{k+1}, and find the remainder on dividing r_{k+1} by the 39-digit number M_{127}. Beginning with $r_1 = 4$, Lucas had to repeat this process of squaring, subtracting 2, and finding the remainder 125 times, and all without making a mistake! Yet I can sit at my computer, apply the Lucas–Lehmer test using the symbolic language *Maple*, and confirm that both M_{31} and M_{127} are primes in an instant.

In the last few pages we have been concentrating on Mersenne primes. To keep our understanding of the prime numbers in perspective, it may be helpful to make the following points:

- The Mersenne numbers $M_n = 2^n - 1$, where n is any positive integer, are very sparse.

- The Mersenne numbers $M_n = 2^n - 1$, where n is a *prime* number, are more sparse than those for which n is any positive integer.

- The Mersenne *primes*, that is the Mersenne numbers M_n that are themselves prime, are even sparser than those for which n is prime.

- The number of primes smaller than n behaves like $n/\log n$, for large n. Thus, compared to the Mersenne primes, the prime numbers are extremely numerous. Indeed, on the numerical evidence available, prime *pairs* are much more plentiful than Mersenne primes.

- It can be shown (see Problem 4.3.6) that

$$\frac{n}{\log n} > \sqrt{n} \quad \text{for all } n \geq 2,$$

and thus, for n large, the prime numbers are asymptotically at least as numerous as the squares.

To emphasize the points made above, let us look at Table 4.9, which displays all the primes between 450,000 and 450,500. There are 47 primes in this interval, including two prime pairs. There is only one square that lies between 450,000 and 450,500, and there is no Mersenne number in this interval.

450,001	450,011	450,019	450,029	450,067	450,071
450,077	450,083	450,101	450,103	450,113	450,127
450,137	450,161	450,169	450,193	450,199	450,209
450,217	450,223	450,227	450,239	450,257	450,259
450,277	450,287	450,293	450,299	450,301	450,311
450,343	450,349	450,361	450,367	450,377	450,383
450,391	450,403	450,413	450,421	450,431	450,451
450,473	450,479	450,481	450,487	450,493	

TABLE 4.9. The 47 primes between 450,000 and 450,500.

Let us look in more detail at part of the range of numbers between 450,000 and 450,500, by giving the complete factorization of the block of 20 numbers that begins with 450,241, the only square in this range.

$$
\begin{aligned}
450{,}241 &= 11^2 \times 61^2 & 450{,}251 &= 89 \times 5059 \\
450{,}242 &= 2 \times 13 \times 17{,}317 & 450{,}252 &= 2^2 \times 3^3 \times 11 \times 379 \\
450{,}243 &= 3^2 \times 19 \times 2633 & 450{,}253 &= 37 \times 43 \times 283 \\
450{,}244 &= 2^2 \times 31 \times 3631 & 450{,}254 &= 2 \times 7 \times 29 \times 1109 \\
450{,}245 &= 5 \times 17 \times 5297 & 450{,}255 &= 3 \times 5 \times 13 \times 2309 \\
450{,}246 &= 2 \times 3 \times 75{,}041 & 450{,}256 &= 2^4 \times 107 \times 263 \\
450{,}247 &= 7 \times 131 \times 491 & 450{,}257 &= 450{,}257 \\
450{,}248 &= 2^3 \times 23 \times 2447 & 450{,}258 &= 2 \times 3 \times 101 \times 743 \\
450{,}249 &= 3 \times 150{,}083 & 450{,}259 &= 450{,}259 \\
450{,}250 &= 2 \times 5^3 \times 1801 & 450{,}260 &= 2^2 \times 5 \times 47 \times 479
\end{aligned}
$$

TABLE 4.10. Factorization of the numbers between 450,241 and 450,260.

To conclude this section, let us look at a connection between Mersenne primes and *perfect numbers* that was known over two thousand years ago.

Definition 4.3.2 A positive integer n is called perfect if the sum of all its *proper* divisors is equal to n. A proper divisor of n is any divisor that is less than n. ■

The first two perfect numbers are 6 and 28, and we have

$$6 = 1 + 2 + 3 \quad \text{and} \quad 28 = 1 + 2 + 4 + 7 + 14.$$

Definition 4.3.3 Let $\sigma(n)$ denote the sum of *all* positive integers that divide the positive integer n, where σ is the Greek letter *sigma*. ■

I have emphasized the word "all" in the above definition, because $\sigma(n)$ includes n itself in the sum of divisors. Thus the condition for n to be perfect can be expressed in the form $\sigma(n) = 2n$. It follows from Definition 4.3.3 that for any prime p and any positive integer m,

$$\sigma\left(p^m\right) = 1 + p + p^2 + \cdots + p^m = \frac{p^{m+1} - 1}{p - 1}, \qquad (4.28)$$

on using (4.18). In particular, we have $\sigma\left(2^m\right) = 2^{m+1} - 1$.

Theorem 4.3.3 Let q_1 and q_2 denote two distinct primes and let m_1 and m_2 denote any two positive integers. Then we have

$$\sigma\left(q_1^{m_1} q_2^{m_2}\right) = \sigma\left(q_1^{m_1}\right)\sigma\left(q_2^{m_2}\right). \qquad (4.29)$$

We say that the function σ is *multiplicative*.

Proof. Consider the equation

$$\sum_{r=0}^{m_1}\sum_{s=0}^{m_2} q_1^r q_2^s = \sum_{r=0}^{m_1} q_1^r \sum_{s=0}^{m_2} q_2^s. \qquad (4.30)$$

On multiplying out the product of the two sums on the right of (4.30) we have the sum of all terms of the form $q_1^r q_2^s$, where r takes all values from 0 to m_1 and s takes all values from 0 to m_2. Thus we do indeed obtain the double sum on the left of (4.30), and this gives (4.29).   ■

It is not difficult to extend Theorem 4.3.3 to show that if

$$N = q_1^{\alpha_1} q_2^{\alpha_2} \cdots q_k^{\alpha_k}, \qquad (4.31)$$

where $q_1, q_2, \ldots, q_k$ are k distinct primes and $\alpha_1, \alpha_2, \ldots, \alpha_k$ are all positive, then

$$\sigma(N) = \sigma\left(q_1^{\alpha_1}\right)\sigma\left(q_2^{\alpha_1}\right)\cdots\sigma\left(q_k^{\alpha_k}\right).$$

It follows that

$$\sigma(N) = \left(\frac{q_1^{\alpha_1+1} - 1}{\alpha_1 - 1}\right)\left(\frac{q_2^{\alpha_2+1} - 1}{\alpha_2 - 1}\right)\cdots\left(\frac{q_k^{\alpha_k+1} - 1}{\alpha_k - 1}\right). \qquad (4.32)$$

The following result on perfect numbers appears in Book IX of Euclid.

Theorem 4.3.4 If $2^n - 1$ is a prime, $2^{n-1}\left(2^n - 1\right)$ is a perfect number.

Proof. We obtain from Theorem 4.3.3 that if $2^n - 1$ is a prime,

$$\sigma\left(2^{n-1}\left(2^n - 1\right)\right) = \sigma\left(2^{n-1}\right)\sigma\left(2^n - 1\right) = \left(2^n - 1\right)\sigma\left(2^n - 1\right), \qquad (4.33)$$

on using (4.28). Since $2^n - 1$ is a prime, we also have

$$\sigma\left(2^n - 1\right) = 1 + \left(2^n - 1\right) = 2^n.$$

Finally, on writing $N = 2^{n-1}(2^n - 1)$, we deduce from (4.33) that

$$\sigma(N) = (2^n - 1)2^n = 2N,$$

and this completes the proof.   ■

This theorem, which has been known since at least the third century BC, involves Mersenne primes, although these are named after a mathematician who flourished in the seventeenth century! Corresponding to every Mersenne prime M_n there is a perfect number $2^{n-1}M_n$. After 6 and 28, the next three perfect numbers are

$$496, \quad 8128, \quad \text{and} \quad 33{,}550{,}336.$$

Euler showed that the perfect numbers based on the Mersenne primes are the only *even* perfect numbers. However, no odd perfect numbers are known. It is generally believed that there are none, but there is currently no proof of this.

Problem 4.3.1 Argue that every Fermat number is either an odd prime or is divisible by an odd prime. Deduce from this and Theorem 4.3.1 that there is an infinite number of primes, giving an alternative to Euclid's proof of Theorem 4.1.1.

Problem 4.3.2 Use the properties of the Fermat numbers mentioned in Problem 4.3.1 to show that the $(n+1)$th prime satisfies the inequality

$$p_{n+1} < f_n = 2^{2^n} + 1.$$

Problem 4.3.3 Note from Theorem 2.2.2, Fermat's little theorem, that for any prime $p \geq 3$,

$$2^{p-1} \equiv 1 \,(\text{mod } p).$$

Deduce that any Mersenne prime M_p has remainder 1 when divided by p.

Problem 4.3.4 Verify that $M_{n+1} = 2M_n + 1$ for all $n \geq 1$.

Problem 4.3.5 Consider the factorizations of the twenty numbers given in Table 4.10. Let us say that $450{,}250 = 2 \times 5^3 \times 1801$ has five prime factors, namely 2, 5 three times, and 1801. Make a table showing how many of the twenty numbers have one, two, three, and so on, prime factors.

Problem 4.3.6 This requires some knowledge of integration. Verify that

$$\frac{1}{t} < \frac{1}{2t^{1/2}} \quad \text{for} \quad t > 4,$$

and deduce from this and (4.2) that

$$\log n - \log 4 = \int_4^n \frac{dt}{t} < \int_4^n \frac{dt}{2t^{1/2}} = n^{1/2} - 4^{1/2}$$

for $n > 4$. Hence show that $\log n < n^{1/2}$ and thus

$$\frac{n}{\log n} > n^{1/2}$$

for all $n > 4$. (A more delicate analysis can be used to show that the latter inequality holds for all $n \geq 2$.)

Problem 4.3.7 Let $r_1 = 3$, and define

$$r_{k+1} = r_k^2 - 2, \quad \text{for } k \geq 1.$$

Observe that r_1 is the Lucas number L_2 and deduce from Problem 3.2.6 that r_k is the Lucas number L_n with $n = 2^k$.

Problem 4.3.8 Let

$$\alpha = \frac{1}{2}\left(\sqrt{2} + \sqrt{6}\right) \quad \text{and} \quad \beta = \frac{1}{2}\left(\sqrt{2} - \sqrt{6}\right),$$

and define

$$r_k = \alpha^{2^k} + \beta^{2^k},$$

for $r \geq 1$. Verify that $\alpha\beta = -1$ and that $r_1 = 4$. Show by mathematical induction that (r_k) satisfies the same recurrence relation as the sequence (r_k) defined in Problem 4.3.7, and so prove that this is the sequence (r_k) used in the Lucas–Lehmer test, defined by (4.27).

Problem 4.3.9 Use (4.31) and (4.32) to evaluate $\sigma(N)$ for $N = 720$ and $N = 5040$.

4.4   The Riemann Hypothesis

Let us define

$$\zeta(s) = \sum_{n=1}^{\infty} \frac{1}{n^s}, \quad \mathbf{Re}(s) > 1, \tag{4.34}$$

where ζ is *zeta*, the sixth letter of the Greek alphabet, and $s = x + iy$ is a complex number. We call $\zeta(s)$ the Riemann zeta function, after G. F. B. Riemann (1826–1866), although there was much interest in $\zeta(s)$ for real values of s long before Riemann's time. The inequality $\mathbf{Re}(s) > 1$ defines the set of values of s for which the infinite series in (4.34) is convergent.

Let us begin with the case in which s is real. Then $\mathbf{Re}(s) > 1$ is equivalent to $s > 1$. When $s = 1$ we obtain the *harmonic* series, which diverges to infinity, as we will now verify. Let us write

$$S_m = \sum_{n=1}^{2^m} \frac{1}{n}, \tag{4.35}$$

so that S_m is the sum of the first 2^m terms of the series on the right side of (4.34) when $s = 1$. We can express S_m in the form

$$S_m = S_0 + (S_1 - S_0) + (S_2 - S_1) + \cdots + (S_m - S_{m-1}), \qquad (4.36)$$

and it is easy to show that $S_0 = 1$, $S_1 - S_0 = \frac{1}{2}$, and $S_2 - S_1 > \frac{1}{2}$. The general term in the sum on the right of (4.36) is

$$S_{j+1} - S_j = \sum_{n=1}^{2^{j+1}} \frac{1}{n} - \sum_{n=1}^{2^j} \frac{1}{n} = \sum_{n=2^j+1}^{2^{j+1}} \frac{1}{n}. \qquad (4.37)$$

Since $2^{j+1} = 2^j + 2^j$, we note that there are 2^j terms in the last sum on the right of (4.37), and observe that each term is greater than or equal to the last term in the sum. We can therefore write

$$S_{j+1} - S_j \geq 2^j \cdot \frac{1}{2^{j+1}} = \frac{1}{2}, \qquad (4.38)$$

and (4.38) holds for $0 \leq j \leq m - 1$. It then follows from (4.36) that

$$S_m \geq 1 + \frac{1}{2}m \qquad (4.39)$$

for all $m \geq 1$. This proves that $S_m \to \infty$ as $m \to \infty$, and thus the infinite series in (4.34) is not defined for $s = 1$.

If we take $s < 1$ in (4.34), we see that

$$\frac{1}{n^s} > \frac{1}{n}.$$

This shows that each term in the series in (4.34), with $s < 1$, is greater than the corresponding term in the harmonic series, which diverges to infinity. Thus the series in (4.34) diverges for all real values of $s \leq 1$.

With a little knowledge of integration, we can compare the zeta function for real values of $s > 1$ with an integral, arguing that

$$\zeta(s) - 1 < \int_1^\infty \frac{dx}{x^s} = \frac{1}{s-1}. \qquad (4.40)$$

Note that this gives the inequality

$$\zeta(s) < 1 + \frac{1}{s-1} = \frac{s}{s-1} \quad \text{for } s > 1. \qquad (4.41)$$

To derive the inequality (4.40) we have to think of the integral as the area under the curve $y = 1/x^s$, and deduce that

$$\frac{1}{n^s} < \int_{n-1}^n \frac{dx}{x^s} \quad \text{for } n \geq 2.$$

Now we need some knowledge of the convergence of series to be able to state that an infinite series of positive terms that is *bounded above*, as is shown by the inequality (4.41), *converges* to a limit. Thus the zeta function $\zeta(s)$ converges, which implies that it has a finite value, for all real values of $s > 1$. Note that it follows from (4.34) and from (4.41) that for $s > 1$, the function $\zeta(s)$ satisfies the inequalities

$$1 < \zeta(s) < 1 + \frac{1}{s-1}, \tag{4.42}$$

and it follows that

$$\zeta(s) \to 1 \quad \text{as} \quad s \to \infty. \tag{4.43}$$

The values of $\zeta(s)$ are known explicitly when s is an even integer. For example,

$$\zeta(2) = \sum_{n=1}^{\infty} \frac{1}{n^2} = \frac{\pi^2}{6} \quad \text{and} \quad \zeta(4) = \sum_{n=1}^{\infty} \frac{1}{n^4} = \frac{\pi^4}{90}.$$

Leonhard Euler found that for any even number $2m$,

$$\zeta(2m) = \sum_{n=1}^{\infty} \frac{1}{n^{2m}} = (-1)^{m-1} \frac{2^{2m-1} B_{2m}}{(2m)!} \pi^{2m}, \tag{4.44}$$

where B_{2m} is called a Bernoulli number, after Jacob Bernoulli (1654–1705). Since $\zeta(2m)$ is positive, it follows from (4.44) that B_{2m} must alternate in sign, the first few being

$$B_2 = \frac{1}{6}, \quad B_4 = -\frac{1}{30}, \quad B_6 = \frac{1}{42}, \quad B_8 = -\frac{1}{30}, \quad B_{10} = \frac{5}{66}, \tag{4.45}$$

and every B_{2m} is a rational number. One of Jacob's brothers, Johann Bernoulli (1667–1748), was also a most influential mathematician, and there were other members of this family who made important contributions to mathematics and science in succeeding generations. As Howard Eves [9] remarks, "One of the most distinguished families in the history of mathematics and science is the Bernoulli family of Switzerland, which from the late seventeenth century on, produced an unusual number of capable mathematicians and scientists."

Even the results that we have just reviewed, that the values of $\zeta(2m)$ are rational multiples of π^{2m}, should be enough to convince us that there is something very special about the zeta function. The following result, due to Leonhard Euler, puts the zeta function at the very heart of number theory, since $\zeta(s)$ is expressed in terms of the building blocks of the positive integers, the prime numbers. Euler was the most outstanding mathematician of his time and made many significant contributions to mathematics. This result alone ensures that his name will never be forgotten.

Theorem 4.4.1 For any value of $s > 1$,

$$\zeta(s) = \sum_{n=1}^{\infty} \frac{1}{n^s} = \prod_{j=1}^{\infty} \left(1 - \frac{1}{p_j^s}\right)^{-1}, \qquad (4.46)$$

where $p_1, p_2, p_3, \ldots$ denote the prime numbers $2, 3, 5, \ldots$.

Proof. The expression on the right of (4.46) denotes the product of all factors of the form $1/(1 - 1/p_j^s)$, where p_j denotes the jth prime. We call this an infinite product, just as we call the original expression for $\zeta(s)$ in (4.34) an infinite sum or infinite series.

Since $0 < 1/p_j \leq \frac{1}{2} < 1$ for all j, we can use (4.19) to express the jth factor of the infinite product in (4.46) as

$$\left(1 - \frac{1}{p_j^s}\right)^{-1} = 1 + \frac{1}{p_j^s} + \frac{1}{p_j^{2s}} + \frac{1}{p_j^{3s}} + \cdots .$$

If we expand every factor in the infinite product in this way, we find that the right side of (4.46) can be written as a sum of terms of the form

$$\frac{1}{p_1^{\alpha_1 s} p_2^{\alpha_2 s} \cdots p_k^{\alpha_k s}} = \frac{1}{n^s},$$

say, where

$$n = p_1^{\alpha_1} p_2^{\alpha_2} \cdots p_k^{\alpha_k}. \qquad (4.47)$$

In (4.47) k is any positive integer, and $\alpha_1 \geq 0, \alpha_2 \geq 0, \ldots, \alpha_k \geq 0$, and every $n > 1$ is uniquely expressed in this way. What I have written above would have served as a proof of (4.46) in Euler's time. However, given that we know that the series for the zeta function converges for $s > 1$, we need to demonstrate that the infinite product also converges for $s > 1$. See, for example, Hardy and Wright [12], who state that this theorem may be regarded as an analytical expression of the fundamental theorem of arithmetic, that every integer $n > 1$ can be expressed uniquely as a product of powers of the primes. ∎

As we have seen, when s is real the series in (4.34) for $\zeta(s)$ is defined only for $s > 1$. As s is increased, with $s > 1$, the value of $\zeta(s)$ decreases, and tends to 1 as s tends to infinity. However, when we allow s to be complex, the zeta function becomes even more interesting and intriguing. I will not justify the statement, made at the beginning of this section, that the series in (4.34) is defined only for $\mathbf{Re}(s) > 1$ when s is complex. We can represent a complex number $s = x + yi$ as a point in the plane, using x and y as coordinates (see Section 6.4). The whole plane is then called the *complex plane*, and a subset of the complex plane such as that defined by $\mathbf{Re}(s) > 1$ is called a *half-plane*.

Another representation was found for the zeta function that involves integrating a certain function around a curve in the complex plane. This gives a function that coincides with the function $\zeta(s)$ defined on the half-plane $\mathbf{Re}(s) > 1$ by the infinite series in (4.34), but extends it in a natural way to all complex values of s except for $s = 1$. (See, for example, Titchmarsh [15].) I have to omit the details, since this would take us far beyond the scope of this book. Nonetheless, I have included this brief section on the zeta function because it is a fascinating episode in the history of mathematics and also because it may encourage you to learn more about complex variable theory in the future.

This extension of the zeta function into the half-plane $\mathbf{Re}(s) \leq 1$, where the infinite series in (4.34) is not valid, led to a most exciting discovery: that the extended function $\zeta(s)$ has zeros! First, it was immediately clear that the zeta function has an infinite number of real zeros. These are called the *trivial* zeros, given by $s = -2, -4, -6, \ldots$. However, the zeta function has also an infinite number of complex zeros; these are called the *nontrivial* zeros. In 1859 Riemann succeeded P. G. L. Dirichlet (1805–1859) in the chair of mathematics at Göttingen, a most prestigious post previously held by Gauss. Around this time Riemann conjectured that the nontrivial zeros of $\zeta(s)$ are infinite in number, and are all of the form $s = x + yi$ with $x = \frac{1}{2}$. This is the famous *Riemann hypothesis*, that the nontrivial zeros of the complex function $\zeta(s)$ all lie on the line $x = \frac{1}{2}$ in the complex plane. In fact, Riemann *proved* that for all zeros $s = x + yi$ of the zeta function with $y \neq 0$, x must lie between 0 and 1. He also found a practical method for determining zeros of $\zeta(s)$, although it was very difficult to carry out such calculations in the precomputer era.

At the International Congress of Mathematicians held in Paris in 1900, David Hilbert presented twenty-three unsolved problems in mathematics that he regarded as being of especial importance, and whose solution he thought would benefit the development of mathematics. In his eighth problem, Hilbert discussed the Riemann hypothesis, and speculated that its resolution would perhaps help settle the Goldbach conjecture and the twin prime conjecture. Although many of Hilbert's twenty-three problems were solved in the twentieth century, the Riemann hypothesis is still not settled at the time of writing, nor is either of the other two conjectures mentioned in Hilbert's eighth problem. A very important advance in pursuing the Riemann hypothesis was made in 1914 by G. H. Hardy, who proved that there is indeed an infinite number of zeros of $\zeta(s)$ of the form $s = \frac{1}{2} + yi$. While this is certainly supportive of Riemann's hypothesis, it still leaves open the possibility that there are zeros of $\zeta(s)$ of the form $s = x + yi$ with $y \neq 0$ and $x \neq \frac{1}{2}$. N. F. H. von Koch had already proved in 1901 that if the Riemann hypothesis is true, the function $\pi(n)$, denoting the number of primes less than n, satisfies an equation of the form

$$\pi(n) = \text{Li}(n) + O\left(n^{1/2} \log n\right), \tag{4.48}$$

where Li(n) is the logarithmic integral, defined in (4.10). I also need to explain that when we write $f(n) = O(g(n))$ we mean that $|f(n)|$ is less than a constant times $g(n)$ for all n. Thus, if the Riemann hypothesis holds, (4.48) gives a measure of how well Li(n) estimates $\pi(n)$.

Let us write $w = \zeta(s) = u + vi$, where $s = x + yi$. Thus a representation of the zeta function involves *four* variables, namely x, y, u, and v, and so we cannot readily exhibit functions of a complex variable, such as the zeta function, graphically. However, we can depict a three-dimensional object on a two-dimensional page, just as a painter can display a landscape on a two-dimensional canvas. Thus, if we write

$$t = |w| = |\zeta(s)|, \quad \text{where} \quad s = x + yi,$$

we could represent the graph of $t = |\zeta(s)|$, where the real variable t is effectively a function of the two real variables x and y, in the same way as the artist displays a landscape. Of course t is only the *modulus* of the zeta function, but $t = |w| = 0$ at all values of s where $w = \zeta(s) = 0$, and so the graph of $t = |\zeta(s)|$ would show all zeros of the zeta function.

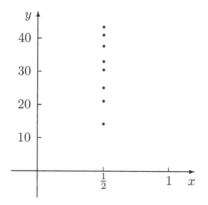

FIGURE 4.3. The first few zeros of $\zeta(s)$ of the form $s = \frac{1}{2} + yi$, with $y > 0$. Corresponding to each zero $\frac{1}{2} + yi$ there is a zero of the form $\frac{1}{2} - yi$.

All of the nontrivial zeros of the zeta function found so far are of the form $\frac{1}{2} + yi$, and an argument based on complex conjugates shows that if $s = x + yi$ is a zero, so is $\bar{s} = x - yi$. The nontrivial zero with the smallest value of $y > 0$ is $\frac{1}{2} + 14.135i$, to three decimal places, and the next few have values of y given by 21.022, 25.011, 30.425, 32.935, 37.586, 40.919, and 43.327, to three decimal places. (See Figure 4.3.) A great advance in the computing of zeros of the zeta function was made by E. C. Titchmarsh, who found over a hundred zeros in 1935 using a punched-card calculating machine. The 100th zero of $\zeta(s)$ is $\frac{1}{2} + 236.524i$, to three decimal places. Finding zeros of the zeta function became easier after the arrival of the automatic digital computer, and now billions of zeros have been found. There are 29 zeros of the form $\frac{1}{2} + yi$ with $0 < y < 10^2$, 649 such zeros with $0 < y < 10^3$, and 1,747,146 zeros for which $0 < y < 10^6$. The distribution of the zeros of $\zeta(s)$ seems as mysterious as that of the primes themselves.

More about the Riemann hypothesis and further interesting material on the prime numbers can be found in the beautifully written book *The Music of the Primes* by Marcus du Sautoy [6].

Problem 4.4.1 Use (4.44) and the values of the Bernoulli numbers given in (4.45) to show that

$$\zeta(6) = \frac{\pi^6}{945}, \quad \zeta(8) = \frac{\pi^8}{9450}, \quad \text{and} \quad \zeta(10) = \frac{\pi^{10}}{93,555}.$$

Problem 4.4.2 Assuming that the conjugate of $\zeta(s)$ is $\zeta(\bar{s})$, deduce that if s is a zero of the zeta function, so is its conjugate $\bar{s}$.

5
Choice and Chance

The excitement that a gambler feels when making a bet is equal to the amount he might win times the probability of winning it.

Blaise Pascal (1623–1662)

5.1 Arrangements and Permutations

This section is concerned with combinatorial mathematics. One of the simplest combinatorial problems is to determine the number of *arrangements* of r objects. Two objects, a_1 and a_2, can be arranged in just two ways, $a_1 a_2$ and $a_2 a_1$, and three objects, a_1, a_2, and a_3, can be arranged in six ways,

$$a_1 a_2 a_3, \quad a_1 a_3 a_2, \quad a_2 a_1 a_3, \quad a_2 a_3 a_1, \quad a_3 a_1 a_2, \quad a_3 a_2 a_1.$$

An *arrangement* is also called a *permutation*. Can you answer the following questions?

1. How many arrangements are there of 4 objects?

2. How many arrangements are there of r objects, for any $r \geq 1$?

I suggest that you put this book down for the moment and see how far you can get in answering these questions before continuing with the rest of this section.

I will now consider a more general problem. Suppose we have r boxes, numbered from 1 up to r, and $n \geq r$ objects, numbered from 1 up to n. In how many ways can we put exactly one of the n objects in each of the r boxes? Let us denote this number by $A_n(r)$. Note that $A_r(r)$ is just the number of ways in which we can arrange r objects in order. Thus, for example, $A_3(3) = 6$, as we found above. We have n choices of how to fill the first box, and if $r > 1$, for each of these choices we can use the remaining $n - 1$ objects to fill the remaining $r - 1$ boxes in $A_{n-1}(r - 1)$ ways. Thus we have

$$A_n(r) = nA_{n-1}(r - 1). \tag{5.1}$$

If $r > 2$, we can similarly express $A_{n-1}(r - 1)$ in terms of $A_{n-2}(r - 2)$ and then use (5.1) to give

$$A_n(r) = n(n - 1)A_{n-2}(r - 2). \tag{5.2}$$

Then we can extend (5.1) and (5.2) to give

$$A_n(r) = n(n - 1) \cdots (n - r + 2)A_{n-r+1}(1), \tag{5.3}$$

and since $A_{n-r+1}(1) = n - r + 1$, we find that

$$A_n(r) = n(n - 1) \cdots (n - r + 2)(n - r + 1) = \frac{n!}{(n - r)!}, \tag{5.4}$$

where $n!$ is n factorial, the product of the first n positive integers, which we encountered in (2.56), the infinite series for e. It is convenient to define $0! = 1$, so that (5.4) holds when $n = r$.

The special case of (5.4) when $n = r$ answers the above question concerning the number of ways of arranging r objects. This is

$$A_r(r) = r!. \tag{5.5}$$

Suppose the n objects are white tennis balls, and we paint r of them red. How many of the $A_n(r)$ arrangements are there such that every box has a *red* ball in it? This is just the same as the number of ways of arranging r balls in r boxes, which is $A_r(r) = r!$.

Now let us write $B_n(r)$ to denote the number of ways of choosing r objects out of n. For example, the different ways of choosing two objects from the set $\{a_1, a_2, a_3, a_4\}$ are

$$a_1a_2, \quad a_1a_3, \quad a_1a_4, \quad a_2a_3, \quad a_2a_4, \quad a_3a_4,$$

and thus $B_4(2) = 6$. We can see from the above discussion that

$$A_n(r) = B_n(r) \times r!. \tag{5.6}$$

Thus the number of ways of putting exactly one of the n objects into each of the r boxes is equal to $r!$ *times* the number of ways of *choosing*

r objects from the n available objects. We will now replace $B_n(r)$ with the standard notation $\binom{n}{r}$, which is read as "n choose r" and is called a *binomial coefficient*. We see from (5.6) and (5.4) that

$$\binom{n}{r} = \frac{n!}{(n-r)!\,r!}. \tag{5.7}$$

The expression on the right of (5.7) is obviously a positive rational number, and from its derivation we know it is a positive integer.

Now, beginning with $n > 1$ white tennis balls, let us paint one of them blue. The number of ways of choosing r balls that *include* the blue ball is just the number of ways of choosing $r-1$ balls from $n-1$, which is $\binom{n-1}{r-1}$. Second, the number of ways of choosing r balls out of the original n that *exclude* the blue ball is just the number of ways of choosing r balls from $n-1$, which is $\binom{n-1}{r}$. Since each choice of r balls must either include or not include the blue ball, we deduce that

$$\binom{n}{r} = \binom{n-1}{r-1} + \binom{n-1}{r}. \tag{5.8}$$

This is called Pascal's identity, which we can check algebraically. We write

$$\binom{n-1}{r-1} + \binom{n-1}{r} = \frac{(n-1)!}{(n-r)!\,(r-1)!} + \frac{(n-1)!}{(n-r-1)!\,r!},$$

and we can simplify this to give

$$\binom{n-1}{r-1} + \binom{n-1}{r} = (n-1)! \cdot \frac{(r+(n-r))}{(n-r)!\,r!}$$
$$= \frac{n!}{(n-r)!\,r!} = \binom{n}{r},$$

which justifies (5.8). In (5.7) we have defined the binomial coefficient for integers $n \geq r \geq 1$. It is useful to extend the definition to include $n = r = 0$ so that the Pascal identity (5.8) holds also for $r = 1$. This means we must define $\binom{n}{0} = 1$, for all $n \geq 0$. Then the binomial coefficients may be written as in Table 5.1, where row $n+1$ contains the numbers $\binom{n}{r}$ for $0 \leq r \leq n$. Each row begins and ends with the integer 1, and it follows from the Pascal identity (5.8) that every *interior* number in any row of the table is just the sum of the two nearest numbers in the row above. The symmetry in the Pascal table is due to the fact that

$$\binom{n}{r} = \binom{n}{n-r}. \tag{5.9}$$

This is easily verified using (5.7). It also follows from the observation that when we choose r objects from n, there are $n-r$ objects remaining, and so a choice of r objects from n is equally a choice of $n-r$ objects from n.

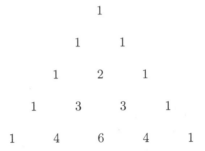

TABLE 5.1. Pascal triangle.

The Pascal table also suggests that for any given value of n, corresponding to a row in the table, the numbers $\binom{n}{r}$ increase with r to a greatest value and then decrease. To verify this, we begin by showing that

$$\binom{n}{r+1} = \frac{n-r}{r+1} \cdot \binom{n}{r},$$

and deduce that

$$\binom{n}{r+1} \geq \binom{n}{r}, \tag{5.10}$$

provided that $n - r \geq r + 1$, or, equivalently, $r \leq \frac{1}{2}(n-1)$. We note also that when $r \geq \frac{1}{2}(n-1)$ we must replace $\geq$ by $\leq$ in (5.10). If n is even, we cannot have $r = \frac{1}{2}(n-1)$, which would give equality in (5.10). In this case we see that the numbers $\binom{n}{r}$ increase from 1 when $r = 0$ to a maximum value, when $r = \frac{1}{2}n$, and then decrease to the value 1 when $r = n$. When n is odd the numbers rise and fall as in the case of n even, except that there are two maximum values that correspond to the two consecutive values $r = \frac{1}{2}(n-1)$ and $r = \frac{1}{2}(n+1)$.

Example 5.1.1 In how many ways can n persons be arranged sitting round a circular table?

We can seat one person arbitrarily, and only need to determine how the others are seated relative to that person, giving $(n-1)!$ arrangements. If n is greater than 2, and we regard two arrangements as being the same if each person has the same two neighbors, we would then have to regard any given arrangement and its mirror image as being equivalent. Thus, for $n > 2$, we would have only $\frac{1}{2}(n-1)!$ allowable arrangements. ∎

Example 5.1.2 In how many ways can we arrange the letters of the word mathematics?

The letters m, a, and t each occur twice, and the remaining five letters, h, e, i, c, and s, occur only once. We can label the two m's as m_1 and m_2, and similarly distinguish the two a's and two t's. There are 11! arrangements of the eleven objects

$$m_1, a_1, t_1, h, e, m_2, a_2, t_2, i, c, s. \tag{5.11}$$

Suppose we cannot see the suffixes that occur in (5.11), so that (5.11) looks like

$$m, a, t, h, e, m, a, t, i, c, s. \tag{5.12}$$

Since there are 2! ways of arranging the two a's, 2! ways of arranging the two a's, and 2! ways of arranging the two t's, there are $2! \times 2! \times 2!$ ways of arranging the m's, a's and t's in (5.11) to give all the arrangements, including (5.11), which look like (5.12) if we cannot see the suffixes. The same argument applies to *any* arrangement of the eleven objects in (5.11). We conclude that there are

$$\frac{11!}{2!\,2!\,2!} = 4{,}989{,}600$$

arrangements of the letters in the word mathematics.   ∎

Example 5.1.3 Suppose we have n distinct nodes arranged arbitrarily around the circumference of a circle. How many chords are obtained by connecting each pair of nodes, and how many points are there in the interior of the circle where two chords intersect?

The number of chords is just the number of ways of choosing 2 from n, which is

$$\binom{n}{2} = \frac{n!}{(n-2)!2!} = \frac{1}{2}n(n-1). \tag{5.13}$$

We obtain an interior point of intersection of two chords by taking any four nodes in order around the circle, say A, B, C, and D, and choosing the point of intersection of AC and BD. Thus the number of interior points obtained by joining every pair of nodes is not greater than the number of ways of choosing four nodes out of n, which is

$$\binom{n}{4} = \frac{n!}{(n-4)!4!} = \frac{1}{24}n(n-1)(n-2)(n-3), \tag{5.14}$$

for $n \geq 4$. Since three or more chords can have a common point, the number of points of intersection can be less than that given by (5.14).   ∎

The problem involving chords discussed in Example 5.1.3 is the basis of the following well-known mathematical curiosity. If we draw all possible chords connecting n nodes arranged on the circumference of a circle, into how many regions is the circle dissected? Let R_n denote the *largest* number of such regions, for any choice of n nodes. An inspection of the diagrams in Figure 5.1 shows that $R_2 = 2$, $R_3 = 4$, $R_4 = 8$, and $R_5 = 16$. If someone who has seen the diagrams in Figure 5.1 is asked to predict the value of R_6, a likely answer is 32. As we will see, this is not true. For, although

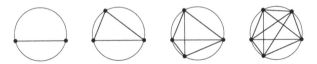

FIGURE 5.1. Chords connecting n nodes on the circumference of a circle.

$R_n = 2^{n-1}$ for $n = 2$, 3, 4, and 5, this formula does not hold for any other value of n, unless we count the case $n = 1$.

Suppose we have $n \geq 4$ nodes. Let C_n denote the number of chords, and let P_n denote the number of points of intersection of the chords in the case in which each pair of chords has a different point of intersection. As we saw in Example 5.1.3,

$$C_n = \binom{n}{2} \quad \text{and} \quad P_n = \binom{n}{4}. \tag{5.15}$$

Of course $P_n = 0$ when $n < 4$. It is obvious that when $n = 4$ or 5, the number of intersection points is equal to P_n for *any* configuration of nodes. However, with $n = 6$, let us choose the nodes as the vertices of a regular hexagon and let us label the six nodes A_1, A_2, A_3, A_4, A_5, and A_6, in order around the circle. (You will find it helpful to draw a diagram.) For this symmetric choice of nodes, the center of the circle is common to the three chords A_1A_4, A_2A_5, and A_3A_6. There are only 13 intersection points in the interior of the circle, which is dissected into 30 regions. You should verify that for all other configurations of 6 nodes, there are $P_6 = 15$ points of intersection, and the circle is dissected into 31 regions.

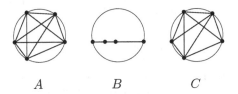

A   B   C

FIGURE 5.2. Removal of one chord.

Now let us determine the value of R_n for any $n \geq 2$. Let us consider how the removal of one chord changes the number of interior points and the number of regions into which the circle is dissected. Diagram A in Figure 5.2 shows the dissection of the circle obtained when there are 5 nodes, diagram B indicates a chord that is to be removed, and diagram C shows the dissection of the circle after the removal of the chosen chord. In this case there are two interior points on the chord that is removed, and the dissected circle in diagram C has 3 fewer regions than the dissected circle in diagram A. In a general configuration, if we remove a chord that contains k interior points of the configuration, we will reduce the number of regions by $k + 1$. Then, given a configuration that has n nodes and the maximum possible number of interior points, which is P_n, let us remove all C_n chords, leaving only one region, the whole circle. Each time a chord is removed, we reduce the number of regions by *one more* than the number of interior points removed. Consequently, when we remove all the chords, we reduce the number of regions by $P_n + C_n$, leaving just one region, so that

$$R_n = P_n + C_n + 1 = \binom{n}{4} + \binom{n}{2} + 1,$$

which we can simplify to give, for all $n \geq 1$,

$$R_n = \frac{1}{24}(n^4 - 6n^3 + 23n^2 - 18n + 24). \tag{5.16}$$

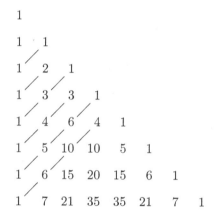

FIGURE 5.3. Fibonacci numbers hidden in the Pascal triangle.

Example 5.1.4 It may come as a surprise to find that the Fibonacci numbers are to be found in the Pascal triangle. For this example it is convenient to realign the numbers in each row in Table 5.1 to give the form in Table 5.3. Then the sum of the numbers in each "rising diagonal," which are shown linked together in Table 5.3, gives a Fibonacci number. For example, we have

$$1 + 5 + 6 + 1 = 13 = F_7 \qquad \text{and} \qquad 1 + 6 + 10 + 4 = 21 = F_8.$$

In general, we have

$$F_{n+1} = \sum_{r=0}^{[n/2]} \binom{n-r}{r}, \tag{5.17}$$

for $n \geq 0$, where $[n/2]$ denotes the greatest integer that is not greater then $n/2$. For example, $[6/2] = 3$, and $[7/2] = 3$. We could say that the sum in (5.17) is over all r such that $0 \leq r \leq n-r$. We can easily verify that (5.17) holds for $n = 0$ and 1. Let us assume that (5.17) holds for $n = k - 1$ and $n = k$, where $k > 1$. Then we have

$$F_k = \sum_{r=0}^{[(k-1)/2]} \binom{k-1-r}{r} \tag{5.18}$$

and

$$F_{k+1} = \sum_{r=0}^{[k/2]} \binom{k-r}{r}. \tag{5.19}$$

Let us put $s = r + 1$ in (5.18) and so obtain

$$F_k = \sum_{s=1}^{[(k-1)/2]+1} \binom{k-s}{s-1}.$$

In the latter summation we can show that $[(k-1)/2]+1 = [(k+1)/2]$, and replace s by r, giving

$$F_k = \sum_{r=1}^{[(k+1)/2]} \binom{k-r}{r-1}. \tag{5.20}$$

We also obtain from (5.19) that

$$F_{k+1} = 1 + \sum_{r=1}^{[k/2]} \binom{k-r}{r}. \tag{5.21}$$

We now use (3.1), the recurrence relation for the Fibonacci numbers, and combine (5.20) and (5.21) to give

$$F_{k+2} = 1 + \sum_{r=1}^{[(k+1)/2]} \binom{k-r}{r-1} + \sum_{r=1}^{[k/2]} \binom{k-r}{r}.$$

Finally, we combine the latter two sums, using Pascal's identity (5.8), checking carefully what happens to the last term in each summation when k is even and when k is odd. We thus find that (5.17) holds for $n = k + 1$, and it follows by mathematical induction that (5.17) holds for all $n \geq 0$.   ■

Example 5.1.5 Let $S_n = \{a_1, a_2, \ldots, a_n\}$, and recall that $\mathcal{P}(S_n)$ denotes the power set of S_n. (See Definition 2.5.5.) We saw in Problem 2.5.3 that $\mathcal{P}(S_n)$ has 2^n elements. The number of subsets of S_n that contain r members is the number of ways of choosing r objects from n, which is $\binom{n}{r}$. Since each subset of S_n contains either $0, 1, 2, \ldots, n-1$, or n members, it follows that

$$\sum_{r=0}^{n} \binom{n}{r} = 2^n, \tag{5.22}$$

and this holds for $n \geq 0$. Thus the sum of the numbers in any row of the Pascal triangle is a power of 2. We will see in the next section that this identity, which we have proved using a combinatorial argument, can also be verified algebraically.   ∎

We conclude this section with a combinatorial interpretation of the Fibonacci numbers. Given any positive number n, let t_n denote the number of ways we can tile a $1 \times n$ rectangle, using 1×1 squares and 1×2 rectangles. I will call a 1×2 rectangle a *domino*. When $n = 4$, we obtain the five tilings that are displayed in Figure 5.4.

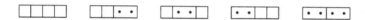

FIGURE 5.4. The five tilings of a 1×4 rectangle by squares and dominoes.

The first of these five tilings consists of four squares, the next three are made up of two squares and a domino, and the last consists of two dominoes. I have placed two spots on each domino to emphasize the difference in its appearance from that of two squares. It is easy to compute t_n when n is small, and the values for $1 \leq n \leq 4$ are given in Table 5.2.

n	1	2	3	4
t_n	1	2	3	5

TABLE 5.2. The number of tilings of a $1 \times n$ rectangle by squares and dominoes.

Every tiling of a $1 \times n$ rectangle begins with a square or a domino. For $n \geq 2$ the number of tilings of a $1 \times n$ rectangle that begin with a square on the left is just t_{n-1}, the number of tilings of a $1 \times (n-1)$ rectangle. The number of tilings of a $1 \times n$ rectangle that begin with a domino on the left is just t_{n-2}, the number of tilings of a $1 \times (n-2)$ rectangle, and this is true for all $n \geq 3$. Thus we have

$$t_n = t_{n-1} + t_{n-2}, \tag{5.23}$$

for $n \geq 3$.

We observe from (5.23) that the sequence (t_n) satisfies the same recurrence relation as the sequence of Fibonacci numbers (F_n). We also note from Table 5.2 and our knowledge of Fibonacci numbers that $t_1 = F_2$ and $t_2 = F_3$ (and also that the next two members of the sequences (t_n) and

(F_n) agree). Let us assume that $t_n = F_{n+1}$ and $t_{n+1} = F_{n+2}$ for some $n \geq 1$, and we note that these equations hold when $n = 1$. On writing

$$t_{n+2} = t_{n+1} + t_n = F_{n+2} + F_{n+1} = F_{n+3},$$

we find that $t_{n+2} = F_{n+3}$. Then, using mathematical induction, we can verify the following theorem.

Theorem 5.1.1 Let t_n denote the number of tilings of a $1 \times n$ rectangle by squares and dominoes. Then we have

$$t_n = F_{n+1}, \tag{5.24}$$

for $n \geq 1$, where F_{n+1} denotes the $(n+1)$th Fibonacci number. ∎

Theorem 5.1.1 tells us that the Fibonacci numbers have a simple combinatorial interpretation, namely that F_n counts the number of ways of tiling a $1 \times (n-1)$ rectangle using squares and dominoes. This is exciting, for it reveals the possibility of justifying Fibonacci identities, or even discovering some, by using combinatorial techniques.

Example 5.1.6 Consider all tilings of a $1 \times 2n$ rectangle. This rectangle consists of $2n$ square cells, which we will label $1, 2, \ldots, 2n$, counting from left to right. For example, in the third of the five tilings displayed in Figure 5.4, cells 2 and 3 are occupied by a domino. Let us now determine how many tilings of a $1 \times 2n$ rectangle have a domino occupying cells n and $n + 1$. Since with $n > 1$, the first $n - 1$ cells and the last $n - 1$ cells can be independently tiled in F_n ways, there are F_n^2 ways of tiling a $1 \times 2n$ rectangle in which a domino occupies cells n and $n+1$. If we do *not* have a domino occupying cells n and $n + 1$, we can split the $1 \times 2n$ rectangle into cells 1 to n and cells $n + 1$ to $2n$, and tile each of these two $1 \times n$ rectangles independently in F_{n+1} ways. Thus there are F_{n+1}^2 ways of tiling a $1 \times 2n$ rectangle that do not have a domino occupying cells n and $n + 1$. Finally, the number of ways of tiling a $1 \times 2n$ rectangle is the sum of the number of tilings in which a domino occupies cells n and $n + 1$ and the number of tilings in which a domino does *not* occupy cells n and $n + 1$. This shows that

$$F_{2n+1} = F_{n+1}^2 + F_n^2, \tag{5.25}$$

for all $n > 1$, which we verified in Problem 3.2.5 using the Binet form (3.12). ∎

The method used in Example 5.1.6 can easily be extended (see Problem 5.1.5) to verify the identity

$$F_{m+n} = F_m F_{n+1} + F_{m-1} F_n.$$

Let $u_{n,r}$ denote the number of tilings of a $1 \times n$ rectangle that contain exactly r dominoes. Of such tilings, $u_{n-1,r}$ begin with a square and $u_{n-2,r-1}$

begin with a domino. Thus

$$u_{n,r} = u_{n-1,r} + u_{n-2,r-1}, \tag{5.26}$$

and we also see that $u_{n,0} = 1$ and $u_{n,r} = 0$ for $r > \frac{1}{2}n$. Since every tiling of a $1 \times n$ rectangle contains r dominoes, where $0 \le r \le [\frac{n}{2}]$, we see that

$$\sum_{r=0}^{[n/2]} u_{n,r} = t_n = F_{n+1}, \tag{5.27}$$

which we can compare with (5.17). Indeed, we infer from (5.26) and the values of $u_{n,0}$ and $u_{n,r}$ for $r > \frac{1}{2}n$ that

$$u_{n,r} = \binom{n-r}{r}, \tag{5.28}$$

and this justifies (5.17). Note that (5.26) is consistent with Pascal's identity (5.8).

For further material on combinatorial methods involving Fibonacci numbers, and an account of similar methods for Lucas numbers, see *Proofs that really count* by Benjamin and Quinn [1].

Problem 5.1.1 In how many ways can 5 black counters, 3 red counters, and 2 white counters be arranged in a row?

Problem 5.1.2 In how many ways can the letters of the word independent be arranged?

Problem 5.1.3 Consider a convex polygon with $n \ge 4$ vertices, and let us define a diagonal as a chord that joins two nonadjacent vertices. How many diagonals does the polygon have?

Problem 5.1.4 Show that R_n, defined by (5.16), can be expressed in the form

$$R_n = \sum_{r=0}^{4} \binom{n-1}{r}.$$

Problem 5.1.5 Consider the F_{m+n} tilings of a $1 \times (m+n-1)$ rectangle using squares and dominoes. Verify that for those tilings in which a domino occupies cells $m-1$ and m, cells 1 to $m-2$ and cells $m+1$ to $m+n-1$ can be independently tiled in F_{m-1} and F_n ways, respectively. Thus show that there are $F_{m-1}F_n$ tilings in which a domino occupies cells $m-1$ and m. Similarly show that there are $F_m F_{n+1}$ tilings in which a domino does not occupy cells $m-1$ and m. Deduce that

$$F_{m+n} = F_m F_{n+1} + F_{m-1} F_n.$$

5.2   Binomial Coefficients

We will show that

$$\binom{n}{r} = \frac{n!}{r!(n-r)!} \tag{5.29}$$

is the coefficient of x^r in the expansion of $(1+x)^n$ in powers of x. Let us write

$$(1+x)^n = c_n(0) + c_n(1)x + \cdots + c_n(n)x^n = \sum_{s=0}^{n} c_n(s)x^s \tag{5.30}$$

and use the obvious identity

$$(1+x)^n = (1+x)(1+x)^{n-1}. \tag{5.31}$$

On replacing $(1+x)^n$ and $(1+x)^{n-1}$ in (5.31) by their expansions in powers of x, as in (5.30), we find that

$$c_n(0) + c_n(1)x + \cdots + c_n(n)x^n = (1+x)\left(c_{n-1}(0) + \cdots + c_{n-1}(n-1)x^{n-1}\right).$$

If we compare the coefficients of x^r on both sides of the above equation, we obtain the recurrence relation

$$c_n(r) = c_{n-1}(r) + c_{n-1}(r-1), \tag{5.32}$$

for $r > 0$. Let us write down the numbers $c_n(r)$ in a triangular formation in Table 5.3, as we did with the binomial coefficients in Table 5.1. It is clear from (5.30) that $c_n(0) = c_n(n) = 1$ for all $n \geq 0$. Thus, for each integer $n \geq 0$,

$$c_n(r) = \binom{n}{r} \quad \text{for} \quad r = 0 \quad \text{and} \quad r = n,$$

and since the numbers $c_n(r)$ and the binomial coefficients are constructed, row by row, for $n = 0, 1, 2, \ldots$, using the same recurrence relation, it follows that

$$c_n(r) = \binom{n}{r} \quad \text{for} \quad n \geq r \geq 0.$$

Then (5.30) may be expressed in the form

$$(1+x)^n = \sum_{s=0}^{n} \binom{n}{s} x^s, \tag{5.33}$$

and this is called the binomial series or the binomial expansion. Note that if we put $x = 1$ in (5.33) we obtain the identity (5.22).

   The derivation of the Pascal identity (5.8) from (5.31) suggests a means of obtaining a more general identity involving the binomial coefficients. Let us begin with

$$(1+x)^n = (1+x)^k(1+x)^{n-k}, \tag{5.34}$$

$$c_0(0)$$

$$c_1(0) \qquad c_1(1)$$

$$c_2(0) \qquad c_2(1) \qquad c_2(2)$$

$$c_3(0) \qquad c_3(1) \qquad c_3(2) \qquad c_3(3)$$

$$c_4(0) \qquad c_4(1) \qquad c_4(2) \qquad c_4(3) \qquad c_4(4)$$

TABLE 5.3. Table of coefficients $c_n(r)$.

from which we recover (5.31) on putting $k = 1$. If we now replace each of the three powers of $(1+x)$ in (5.34) by its binomial series, given by (5.33), we obtain

$$\sum_{s=0}^{n} \binom{n}{s} x^s = \sum_{s=0}^{k} \binom{k}{s} x^s \sum_{s=0}^{n-k} \binom{n-k}{s} x^s. \qquad (5.35)$$

We may assume that $k \leq \frac{1}{2}n$, for otherwise we could replace k by $n - k$, which would leave (5.35) unchanged. Let us now compare the coefficient of x^r on both sides of (5.35), and in view of (5.9) we can choose $r \leq \frac{1}{2}n$. If we expand the right side of (5.35) in powers of x, the number of terms involving x^r depends on whether or not $r \leq k$. There are $r + 1$ terms if $r \leq k$, and there are $k + 1$ terms otherwise. Thus we obtain

$$\binom{n}{r} = \sum_{t=0}^{m} \binom{k}{t} \binom{n-k}{r-t}, \qquad (5.36)$$

where m is the smaller of the two numbers r and k. Note that (5.36) holds for all r and k such that both r and k are not greater than $\frac{1}{2}n$, and (5.36) reduces to the Pascal identity (5.8) when we choose $k = 1$.

Example 5.2.1 In every row of the Pascal triangle the sum of the odd-numbered terms is equal to the sum of the even-numbered terms. Because of the symmetry of the binomial coefficients, given by (5.9), this is obviously true for rows that have an even number of terms. However, this equality also holds in those rows, apart from the first, in which there is an odd number of terms. For example, in rows 3, 5, and 7 we have

$$1 + 1 = 2, \quad 1 + 6 + 1 = 4 + 4, \quad 1 + 15 + 15 + 1 = 6 + 20 + 6.$$

To prove that this holds in general, let us replace n by $2n$ and x by -1 in (5.33) to obtain

$$\sum_{s=0}^{2n} \binom{2n}{s} (-1)^s = 0,$$

where the numbers $(-1)^s$ alternate between $+1$ and -1, beginning with $+1$, corresponding to $s = 0$. If we take all the terms that are multiplied by -1 to the right side of the latter equation, we obtain

$$\sum_{s=0}^{n} \binom{2n}{2s} = \sum_{s=1}^{n} \binom{2n}{2s-1},$$

for $n \geq 1$. ∎

In Chapter 1 we found simple expressions for the sum of the first n positive integers, and also for the sum of the first n triangular numbers. We can now see that these are only the first two of a family of summations involving binomial coefficients. Let us begin with the Pascal identity (5.8), replace n by $s + k$, r by $k + 1$, and rearrange the identity to obtain

$$\binom{s+k}{k+1} - \binom{s+k-1}{k+1} = \binom{s+k-1}{k}. \tag{5.37}$$

Then (5.37) holds for all $s \geq 1$ if we define the second term on the left as zero when $s = 1$. If we sum each term over s, from $s = 1$ to $s = n$, and interchange the left and right sides of the equation, we obtain

$$\sum_{s=1}^{n} \binom{s+k-1}{k} = \sum_{s=1}^{n} \binom{s+k}{k+1} - \sum_{s=2}^{n} \binom{s+k-1}{k+1},$$

since the missing term in the last summation, corresponding to $s = 1$, is defined to be zero. Finally, we observe that all the terms on the right side of this last equation cancel out, except for the last term of the first summation on the right, to give

$$\sum_{s=1}^{n} \binom{s+k-1}{k} = \binom{n+k}{k+1}. \tag{5.38}$$

When $k = 1$ in (5.38), we have the expression obtained in (1.3), showing that the sum of the first n integers is the nth triangular number. On choosing $k = 2$ in (5.38) we have the result obtained in (1.8), that the sum of the first n triangular numbers is the nth tetrahedral number. We saw in Chapter 1 that the names of these two sets of numbers (triangular and tetrahedral) arose from the way they can be depicted geometrically. This prompts the thought that we might also think of the numbers defined by (5.38) for $k > 2$ in a geometrical way if we were not constrained by our everyday experience of living in a world of three dimensions. Indeed, mathematicians do have a name for an n-dimensional object that corresponds to the triangle when $n = 2$ and to the tetrahedron when $n = 3$; it is called the n-dimensional *simplex*.

Conway and Guy [3] give an ingenious verification of (1.8), based on the following equation involving triangular arrays:

$$
\begin{array}{llll}
\begin{array}{l}
1 \\
1\ 2 \\
1\ 2\ 3 \\
1\ 2\ 3\ 4 \\
1\ 2\ 3\ 4\ 5
\end{array}
\quad + \quad
\begin{array}{l}
1 \\
2\ 1 \\
3\ 2\ 1 \\
4\ 3\ 2\ 1 \\
5\ 4\ 3\ 2\ 1
\end{array}
\quad + \quad
\begin{array}{l}
5 \\
4\ 4 \\
3\ 3\ 3 \\
2\ 2\ 2\ 2 \\
1\ 1\ 1\ 1\ 1
\end{array}
\quad = \quad
\begin{array}{l}
7 \\
7\ 7 \\
7\ 7\ 7 \\
7\ 7\ 7\ 7 \\
7\ 7\ 7\ 7\ 7
\end{array}
\end{array}
$$

Each number in the array on the right of the above equation is the sum of the numbers in the corresponding positions in the three arrays on the left of the equation. The second triangular array above is obtained by rotating the first array clockwise through an angle of $2\pi/3$, and the third array is obtained by rotating the second array clockwise through an angle of $2\pi/3$. The sum of the numbers in the rth row of the first array is T_r, the rth triangular number, and the sum of the numbers in the rth row of the triangular array on the right of the equation is just 7 times r. Thus, on summing up all the numbers in each of the four arrays, we find that

$$3(T_1 + T_2 + T_3 + T_4 + T_5) = (1 + 2 + 3 + 4 + 5) \times 7.$$

If we had n rows in each array rather than 5, we would obtain

$$3(T_1 + T_2 + \cdots + T_n) = T_n \times (n + 2) = \frac{1}{2}n(n + 1)(n + 2),$$

and hence

$$T_1 + T_2 + \cdots + T_n = \frac{1}{6}n(n + 1)(n + 2).$$

In (5.33) we have shown how $(1+x)^n$ can be expressed as a series involving multiples of powers of x, and this is valid when n is any integer greater than or equal to zero. In these cases, the series is finite, and $(1 + x)^n$ is a polynomial in x of degree n. What meaning can we assign to $(1+x)^\alpha$ when $1 + x \geq 0$ and α is any real number? We define $(1 + x)^0 = 1$, and we next consider positive values of α. If

$$(1 + x)^\alpha = c_0 + c_1 x + c_2 x^2 + c_3 x^3 + \cdots,$$

we see, by putting $x = 0$, that $c_0 = 1$. If α is an irrational number, we can find rational numbers above and below α, as close to α as we wish. For example, for the irrational number $\sqrt{2}$, we can write

$$1.41421356 < \sqrt{2} < 1.41421357.$$

If we can define $(1+x)^\alpha$ when α is a positive rational number, its definition can be extended to irrational values of α by requiring that $(1 + x)^\alpha$ lie between $(1 + x)^{\alpha_1}$ and $(1 + x)^{\alpha_2}$, where α_1 and α_2 are *any* two rational numbers that lie on either side of α. We achieve this extension of the definition of $(1 + x)^\alpha$ from rational to irrational values of α by requiring that for each value of x, $(1+x)^\alpha$ be a *continuous* function of α.

If α is a positive rational number, we can write $\alpha = m/n$, where m and n are positive integers. Then, for $1 + x \geq 0$, we define

$$(1+x)^{m/n} = y,$$

where y is the unique nonnegative number such that

$$(1+x)^m = y^n.$$

Also, for any positive real number α, we define

$$(1+x)^{-\alpha} = 1/(1+x)^{\alpha}.$$

We need to extend the definition of binomial coefficient. We write

$$\binom{\alpha}{r} = \frac{\alpha(\alpha - 1) \cdots (\alpha - r + 1)}{r!} \tag{5.39}$$

for all real values of α, and all integers $r \geq 1$, and we replace the expression on the right of (5.39) by 1 when $r = 0$. If α is a nonnegative integer, this definition coincides with the definition of a binomial coefficient that is given in (5.29). Then the binomial series, defined for all x such that $1 + x \geq 0$ and all real α, is

$$(1+x)^{\alpha} = \sum_{r=0}^{\infty} \binom{\alpha}{r} x^r. \tag{5.40}$$

Note that in contrast to the case in which $\alpha = n$ is a nonnegative integer, the series on the right of (5.40) is infinite. I will not justify this result here. It can be verified using the theory of Taylor series. The series on the right of (5.40) is valid for $-1 < x < 1$. For then the infinite series is convergent and thus has a meaningful sum.

Example 5.2.2 Let us use (5.39) to evaluate the first few binomial coefficients when $\alpha = \frac{1}{2}$. We have $\binom{1/2}{0} = 1$, $\binom{1/2}{1} = \frac{1}{2}$,

$$\binom{1/2}{2} = \frac{\frac{1}{2}(\frac{1}{2} - 1)}{2!} = -\frac{1}{8},$$

$$\binom{1/2}{3} = \frac{\frac{1}{2}(\frac{1}{2} - 1)(\frac{1}{2} - 2)}{3!} = \frac{1}{16},$$

and

$$\binom{1/2}{4} = \frac{\frac{1}{2}(\frac{1}{2} - 1)(\frac{1}{2} - 2)(\frac{1}{2} - 3)}{4!} = -\frac{5}{128}.$$

These coefficients alternate in sign, apart from the first two, and it is clear that $\binom{1/2}{r}$ has the sign $(-1)^{r+1}$ for all $r \geq 1$.   ∎

We will now look at some interesting special cases of (5.40), beginning with $(1 + x)^{-1}$. Let us write

$$(1 + x)^{-1} = c_0 + c_1 x + c_2 x^2 + c_3 x^3 + \cdots,$$

and multiply both sides of the latter equation by $1 + x$ to obtain

$$1 = (1 + x)(c_0 + c_1 x + c_2 x^2 + c_3 x^3 + \cdots). \tag{5.41}$$

If we put $x = 0$ in (5.41), we find that $c_0 = 1$. If we equate coefficients of x^n in (5.41) for $n > 0$, we obtain

$$0 = c_{n-1} + c_n,$$

and since $c_0 = 1$, it follows that $c_n = (-1)^n$, for all $n \geq 0$, giving

$$(1 + x)^{-1} = 1 - x + x^2 - x^3 + \cdots.$$

This expansion is valid for $-1 < x < 1$. If we replace x by $-x$, we obtain

$$(1 - x)^{-1} = 1 + x + x^2 + x^3 + \cdots, \qquad -1 < x < 1, \tag{5.42}$$

as we found in (4.19). We can extend the above result to show by induction that

$$(1 - x)^{-m-1} = \sum_{r=0}^{\infty} \binom{m + r}{m} x^r, \qquad -1 < x < 1. \tag{5.43}$$

We see from (5.42) that this is valid for $m = 0$. Let us assume that (5.43) holds for $m = k \geq 0$. Then we can write

$$(1 - x)^{-k-2} = (1 - x)^{-1}(1 - x)^{-k-1} = (1 + x + x^2 + \cdots) \sum_{r=0}^{\infty} \binom{k + r}{k} x^r.$$

Then the coefficient of x^n in the expansion of $(1 - x)^{-k-2}$ is

$$\sum_{r=0}^{n} \binom{k + r}{k} = \binom{n + k + 1}{k + 1},$$

on using (5.38) with n replaced by $n + 1$. Thus, by induction, (5.43) holds for all integers $m \geq 0$. The last example we will consider is the expansion of $(1 + x)^{1/2}$. Let us write

$$(1 + x)^{1/2} = 1 + c_1 x + c_2 x^2 + c_3 x^3 + \cdots.$$

Since $(1 + x)^{1/2}(1 + x)^{1/2} = 1 + x$, we have

$$1 + x = (1 + c_1 x + c_2 x^2 + c_3 x^3 + \cdots)(1 + c_1 x + c_2 x^2 + c_3 x^3 + \cdots),$$

and we obtain, on equating coefficients of x, x^2, x^3, and x^4,

$$2c_1 = 1,$$
$$c_1^2 + 2c_2 = 0,$$
$$2c_1c_2 + 2c_3 = 0,$$
$$2c_1c_3 + c_2^2 + 2c_4 = 0.$$

We can determine the value $c_1 = \frac{1}{2}$ from the first of these equations, and then find c_2 from the second equation, c_3 from the third, and c_4 from the fourth. In principle we could in this way determine any number of the coefficients of this series. Thus we find that

$$(1+x)^{1/2} = 1 + \frac{1}{2}x - \frac{1}{8}x^2 + \frac{1}{16}x^3 - \frac{5}{128}x^4 + \cdots, \qquad (5.44)$$

which is consistent with what we found in Example 5.2.2. The above binomial series for $(1+x)^{1/2}$ was first derived by Henry Briggs (1556–1630), the creator of one of the first logarithm tables. Briggs was the first to find a binomial series (5.40) for a value of α that is not an integer.

Example 5.2.3 Let us approximate $\sqrt{2}$ using the binomial expansion. We *could* use (5.44) directly with $x = 1$, to obtain

$$\sqrt{2} = (1+1)^{1/2} \approx 1 + \frac{1}{2} - \frac{1}{8} + \frac{1}{16} - \frac{5}{128} = 1.398,$$

to three decimal places. It is not surprising that this is a very poor approximation to $\sqrt{2}$, since we are using the first few terms of a series that converges very slowly. With a little more ingenuity, we note that the square of $7\sqrt{2}$ is 98, which is close to 100. (This approximation was used by Isaac Newton (1642–1727) in a calculation related to the computation of a table of logarithms.) We write

$$\frac{98}{100} = 1 - \frac{2}{100},$$

take square roots, and use the first five terms of the binomial expansion to obtain

$$\frac{7}{10}\sqrt{2} = \left(1 - \frac{2}{100}\right)^{1/2} \approx 0.98994949375,$$

and hence obtain $\sqrt{2} \approx 1.4142135625$. The true value of $\sqrt{2}$ is approximately 1.4142135624, correct to ten decimal places. ∎

The method used in Example 5.2.3 for approximating $\sqrt{2}$ can easily be generalized to estimate $\sqrt{m}$, where m is any positive integer that is not a square. We choose positive integers p and q such that $q^2m = p^2 + c$, where $|c|$ is smaller than p^2, so that

$$m = \frac{p^2 + c}{q^2} = \frac{p^2}{q^2}\left(1 + \frac{c}{p^2}\right).$$

We then derive

$$\sqrt{m} = \frac{p}{q}\left(1 + \frac{c}{p^2}\right)^{1/2}, \tag{5.45}$$

and we can use the binomial expansion (5.44) to estimate the square root on the right of (5.45). Let x_0 denote the approximation to $\sqrt{m}$ obtained from (5.45) by using the first two terms of the binomial series (5.44). This yields the approximation

$$\sqrt{m} \approx \frac{p}{q}\left(1 + \frac{c}{2p^2}\right) = x_0. \tag{5.46}$$

Then if we define x_1 as

$$x_1 = \frac{1}{2}\left(x_0 + \frac{m}{x_0}\right), \tag{5.47}$$

we may write

$$x_1 - \sqrt{m} = \frac{1}{2}\left(x_0 + \frac{m}{x_0}\right) - \sqrt{m} = \frac{1}{2x_0}\left(x_0^2 - 2x_0\sqrt{m} + m\right),$$

and thus

$$x_1 - \sqrt{m} = \frac{(x_0 - \sqrt{m})^2}{2x_0}. \tag{5.48}$$

Since $x_0 > 0$, it is clear from (5.48) that $x_1 > \sqrt{m}$, and that if x_0 is sufficiently close to $\sqrt{m}$, then x_1 will be very much closer to $\sqrt{m}$. If necessary, we can repeat this process several times, defining $x_2 = \frac{1}{2}(x_1 + m/x_1)$, and so on. This procedure defined by (5.47) is called Heron's method, after Heron of Alexandria (first century AD). Obviously $\sqrt{m}$ is a solution of the equation $x^2 - m = 0$. The above square root process is also called Newton's method, since it is a special case of Isaac Newton's method for approximating a solution of a general equation in one variable.

Problem 5.2.1 Verify that

$$\sqrt{5} = \frac{9}{4}\left(1 - \frac{1}{81}\right)^{1/2}$$

and, using the first two terms of the binomial expansion (5.44), show that

$$\sqrt{5} \approx \frac{161}{72} \approx 2.23611.$$

Also find the approximation obtained by using the first five terms of the binomial expansion. Check the accuracy of both results.

Problem 5.2.2 Verify that $161/72$, the approximation for $\sqrt{5}$ obtained in Problem 5.2.1, is the fourth convergent of $[2, 4, 4, 4, \ldots]$, the simple continued fraction for $\sqrt{5}$.

Problem 5.2.3 Show that

$$\sqrt{3} = \frac{7}{4}\left(1 - \frac{1}{49}\right)^{1/2},$$

and use the first two terms of the binomial expansion (5.44) to obtain

$$\sqrt{3} \approx \frac{97}{56} = x_0 \approx 1.73214.$$

Compute $x_1 = \frac{1}{2}(x_0 + 3/x_0)$ and determine how close x_0 and x_1 are to $\sqrt{3}$.

5.3   Probability

It has been said that the study of probability theory was motivated by gamblers who wished to know the true odds governing the outcomes of events on which they were betting. It is also said that incurable gamblers will bet on anything, and so there are limits to what a mathematician can do to help a gambler. The mathematician may be of little help when it comes to predicting which of two flies will fly away first, but can offer advice when it comes to more predictable events, for example those that involve tossing coins, rolling dice, or playing cards. I am talking about professional advice. If the gambler should say, "Please advise me, *should* I gamble?" this mathematician would say "No!"

The tossing of a coin is one of the simplest probabilistic activities that is amenable to mathematical analysis. If a coin is tossed 10 times and it comes down tails every time, we cannot be certain that it is biased, although we will have doubts. We assume that we have a coin that, when tossed comes down "heads" or "tails" with the same likelihood. Such a coin is said to be fair, or unbiased. What we are saying is that if we were able to toss the coin an infinite number of times, and h_n denotes the number of heads obtained after n tosses, the limit of the ratio h_n/n tends to $\frac{1}{2}$ as n tends to infinity. (In the notation and language of Section 4.2, this is the same as writing $h_n/n \sim \frac{1}{2}$ and saying that h_n/n is asymptotically equal to $\frac{1}{2}$.) Of course, this experiment cannot be carried out, and so the concept of a fair coin is only an abstraction.

We say that when we toss a coin, the probability of a head is $\frac{1}{2}$ and the probability of a tail is also $\frac{1}{2}$. There is no other possible outcome, and the sum of these two probabilities is 1. If we toss the coin twice, there are four possible outcomes, which are equally likely. In an obvious notation, the four outcomes are

$$HH, \quad HT, \quad TH, \quad TT.$$

The probability of two heads is $\frac{1}{4}$, as is the probability of two tails, and the probability of one head and one tail is $\frac{1}{2}$. Of course, the probability of having one head *followed* by one tail, or vice versa, is $\frac{1}{4}$. If we toss the coin three times, the possible outcomes are

$$HHH, \quad HHT, \quad HTH, \quad HTT, \quad THH, \quad THT, \quad TTH, \quad TTT.$$

These outcomes are all equally likely. Thus the probability of two heads and one tail is the number of times such an outcome occurs, which is 3, divided by 8, the total number of outcomes, giving a probability of $\frac{3}{8}$. The probability of obtaining three heads (or three tails) is $\frac{1}{8}$.

If we toss the coin n times, what is the probability of obtaining exactly r heads? The number of possible outcomes when we toss a coin n times is 2^n, and the number of ways of arranging r copies of the letter H and $n - r$ copies of the letter T is $\binom{n}{r}$. Thus the probability of obtaining exactly r heads is $\binom{n}{r}/2^n$.

Example 5.3.1 The probability of obtaining exactly r heads in ten tosses of a coin is given by

$$p(r) = \frac{1}{2^{10}} \binom{10}{r} = \frac{1}{1024} \binom{10}{r}.$$

It follows from the symmetric property of the binomial coefficients, given in (5.9), that $p(r) = p(10 - r)$. We can also see this by using a combinatorial argument. The probability of obtaining $10 - r$ heads, $p(10 - r)$, is the same as the probability of obtaining r tails. This is equal to the probability of obtaining r heads, which is $p(r)$. For $0 \leq r \leq 5$, the binomial coefficients $\binom{10}{r}$ are 1, 10, 45, 120, 210, and 252, and Table 5.4 gives the values of the probabilities $p(r)$, rounded to three decimal places. There is a lot of information in Table 5.4. For example, we can see that the probability of having at least four heads or four tails in ten tosses is

$$p(4) + p(5) + p(6) \approx 0.205 + 0.246 + 0.205 = 0.656. \quad \blacksquare$$

r	0	1	2	3	4	5
$p(r)$	0.001	0.010	0.044	0.117	0.205	0.246

TABLE 5.4. The probabilities of having r heads in ten tosses.

Example 5.3.2 A die (whose plural is dice) is a cube with one of the numbers 1 to 6 marked on each face. What is the probability of throwing a total of 7 with two dice, and what is the probability of throwing 5 or less? With each throw we obtain 1, 2, 3, 4, 5, or 6. If we throw two dice, there are $6 \times 6 = 36$ possible outcomes, and we achieve 7 from the six outcomes

$$1 + 6, \quad 2 + 5, \quad 3 + 4, \quad 4 + 3, \quad 5 + 2, \quad 6 + 1.$$

Thus the probability of throwing a total of 7 is $\frac{6}{36} = \frac{1}{6}$. You can check that there are ten outcomes that yield 5 or less, and thus the probability of throwing 5 or less is $\frac{10}{36} = \frac{5}{18}$. ■

An ordinary pack of cards consists of 52 cards arranged in four suits, each containing 13 cards. The suits are called spades, hearts, diamonds, and clubs. Each suit contains four face cards, called ace, king, queen, and jack, in diminishing order of importance. These are all more important than the nine nonface cards, which are numbered from ten down to two, again in diminishing order of importance.

Example 5.3.3 If one is dealt four cards, what is the probability that they are all face cards? What are the probabilities that there are exactly three face cards, exactly two, and exactly one? Finally, what is the probability of being dealt no face cards?

There are $4 \times 4 = 16$ face cards. Therefore the number of ways of choosing four face cards is

$$N_4 = \binom{16}{4} = \frac{16!}{4!\,12!}.$$

The number of ways of choosing any four cards is

$$N = \binom{52}{4} = \frac{52!}{4!\,48!}.$$

Thus the probability that the four cards are all face cards is

$$\frac{N_4}{N} = \frac{16!}{4!\,12!} \times \frac{4!\,48!}{52!} = \frac{4}{595} = p_4,$$

say. The number of ways of obtaining three face cards is the number of ways of choosing three cards from the 16 face cards times the number of ways of choosing one card out of the 36 nonface cards. Thus the number of ways of obtaining three face cards is

$$N_3 = \binom{16}{3} \times 36 = \frac{16!}{3!\,13!} \times 36.$$

Thus the probability of obtaining three face cards is

$$\frac{N_3}{N} = \frac{36 \times 16!}{3!\,13!} \times \frac{4!\,48!}{52!} = \frac{576}{7735} = p_3.$$

The number of ways of obtaining two face cards is the number of ways of choosing two face cards from the 16 face cards times the number of ways of choosing two cards from the 36 nonface cards. Thus the number of ways of obtaining two face cards is

$$N_2 = \binom{16}{2} \times \binom{36}{2} = \frac{16!}{2!\,14!} \times \frac{36!}{2!\,34!}.$$

Thus the probability of obtaining two face cards is

$$\frac{N_2}{N} = \frac{16!}{2!\,14!} \times \frac{36!}{2!\,34!} \times \frac{4!\,48!}{52!} = \frac{432}{1547} = p_2.$$

The number of ways of obtaining one face card is the number of ways of choosing one from the 16 face cards times the number of ways of choosing three cards from the 36 nonface cards. Thus the number of ways of obtaining one face card is

$$N_1 = 16 \times \binom{36}{3} = \frac{16 \times 36!}{3!\,33!},$$

and the probability of obtaining one face card is

$$\frac{N_1}{N} = \frac{16 \times 36!}{3!\,33!} \times \frac{4!\,48!}{52!} = \frac{192}{455} = p_1.$$

Finally, the number of ways of choosing four cards from the 36 nonface cards is

$$N_0 = \binom{36}{4} = \frac{36!}{4!\,32!},$$

and so the probability of obtaining no face cards is

$$\frac{N_0}{N} = \frac{36!}{4!\,32!} \times \frac{4!\,48!}{52!} = \frac{99}{455} = p_0.$$

These probabilities are expressed in decimal form in Table 5.5 to make it easier to compare them.

n	0	1	2	3	4
p_n	0.218	0.422	0.279	0.074	0.007

TABLE 5.5. The probabilities of obtaining n face cards, on being dealt four cards.

As a check on these results, note that we must have

$$N_0 + N_1 + N_2 + N_3 + N_4 = N,$$

and on dividing the above equation throughout by N, we obtain

$$p_0 + p_1 + p_2 + p_3 + p_4 = 1,$$

as we should expect. The numbers in Table 5.5 agree with the latter equation. ∎

Example 5.3.4 I conclude this chapter with a problem whose solution most people find surprising. To avoid complicating matters, let us forget about the existence of leap years. Suppose we have N persons whose birthdays we don't know. The number of choices of birthday for the first person is 365. The number of choices of birthday for the first two persons is 365^2, and the number of choices of birthday for all N persons is 365^N. On the other hand, the number of choices of birthday for persons 1 and 2 if they must have different birthdays is 365×364. If all N must have different birthdays, then obviously N cannot exceed 365, and the number of choices if all N are to have different birthdays is

$$365 \times 364 \times 363 \times \cdots \times (365 - N + 1). \qquad (5.49)$$

Thus the probability that all N persons have different birthdays is just the number in (5.49) divided by 365^N, giving

$$\left(1 - \frac{1}{365}\right)\left(1 - \frac{2}{365}\right) \cdots \left(1 - \frac{N-1}{365}\right) = p(N),$$

say, where $2 \leq N \leq 365$. It is clear that $p(N)$ decreases as we increase N. We find that $p(22) \approx 0.524$ and $p(23) \approx 0.493$. Thus in a gathering of 23 people there is a better than fifty-fifty chance of two or more sharing the same birthday.   ∎

Problem 5.3.1 If we toss a coin three times, what is the probability that the three outcomes are all the same, that is, we obtain three heads or three tails? What is the probability that we have two of one kind and one of the other?

Problem 5.3.2 If we throw two dice, what is the probability that we score 7 or more? What is the probability that the score is a prime number, that is, one of the numbers 2, 3, 5, 7, or 11?

6

Geometrical Constructions

Let no one enter who does not know geometry.

Plato (427–347 BC)

Written above the door of Plato's Academy in Athens.

6.1 Ruler and Compasses

Euclid circa 300 BC gave a systematic account of the geometry known at
the time, beginning with certain basic concepts, known as *axioms*. These
axioms are to be thought of as initial assumptions on which the geometry
of Euclid depend. His axioms are as follows:

1. *A straight line may be drawn from any point to any other point.*

2. *A finite straight line may be extended continuously in a straight line.*

3. *A circle may be drawn with any center and any radius.*

4. *All right angles are equal to one another.*

5. *If a straight line meets any two other straight lines so as to make
   the two interior angles on one side of it together less than two right
   angles, the other straight lines, if extended indefinitely, will meet on
   that side on which the angles are less than two right angles.*

At the heart of ancient Greek geometry are the "ruler and compass" constructions Two implements were used, a *ruler* for drawing straight lines and a *pair of compasses* for drawing circles. The ruler is simply a straight edge with no markings on it. It is used only for drawing straight lines, and not for measuring lengths. The compasses consist of two arms connected by a movable joint. At the end of one arm there is a sharp point that is placed at the center of the circle to be drawn. There is a a a pencil at the end of the other arm, which can be moved to change the radius of the circle. Set against their many successes in this area of geometry, there were a few constructions by ruler and compass that the Greeks kept vainly struggling to achieve. After more than two millennia of experience of ruler and compass constructions, mathematicians at last attained a fuller understanding of the limitations of these methods and were able to *prove* that these long-standing unsolved classical problems were truly unsolvable. However, it is not difficult, after a little experimenting, to rediscover for ourselves some of the more obvious constructions that *can* be carried out by ruler and compasses.

Construction 1 *Draw a line that is perpendicular to a given line at a given point A.*

We use the compasses to mark two points B and C on the given line, equally spaced on either side of A. See Figure 6.1. We then draw equal arcs centered at B and C and label their point of intersection D. Then AD meets BC at right angles. To see this, note that in the triangles ADB and ADC, $|AB| = |AC|$, $|DB| = |DC|$, and DA is common to both triangles. (We write $|AB|$ to denote the *length* of the line segment AB.) Thus the two triangles are congruent. Finally, the angles DAB and DAC must both be right angles, since they are equal and their sum is equal to two right angles.   ■

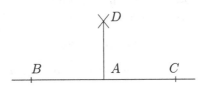

FIGURE 6.1. Draw a line that is perpendicular to the horizontal line at A.

Construction 2 *Draw a line through a given point A that is perpendicular to a given line that does not pass through A.*

With center A we use the compasses to mark off two points, B and C, on the line. (Draw a diagram.) Then we draw equal arcs centered at B and C to meet at a point D that is on the other side of the given line from A.

Let AD and BC intersect at the point E. We deduce that the triangles ABD and ACD are congruent, and that triangles ABE and ACE are congruent. Thus the angles AEB and AEC are equal and so must both be right angles. ∎

Construction 3 *Find the midpoint of a given line segment BC.*

This construction is similar to Construction 2. With any radius greater than half the distance BC we draw arcs of the same length, centered at B and C, to intersect at a point A on one side of BC and at a point D on the other side of BC. Then we can verify that E, the point where AD and BC intersect, is the midpoint of BC, and that AE is the *perpendicular bisector* of BC.   ∎

Construction 4 *Draw a line through a given point A that is parallel to a given line.*

We begin by using Construction 2 to obtain points B and C on the given line and the midpoint D of BC. Then AD is perpendicular to BC. Next, using the ruler, we extend the line DA and use the compasses to find a point E on the extended line DA such that A is the midpoint of DE. Finally we construct the perpendicular bisector of the line DE. This line passes through A and, being perpendicular to DE, must be parallel to the original line BC.   ∎

Construction 5 *Given two lengths a and b, with a > b, construct the lengths a + b and a − b.*

These constructions are very simple. We use the compasses to mark off the lengths $|OA| = a$ and $|AB| = b$, as shown in the diagram on the left of Figure 6.2. Then the length $|OB|$ equals $a + b$. The construction of $a - b$ is shown in the diagram on the right of Figure 6.2.   ∎

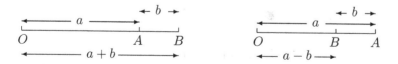

FIGURE 6.2. Construction of $a + b$ and $a - b$.

Construction 6 *Given two lengths a and b and a unit length, construct the lengths ab and a/b.*

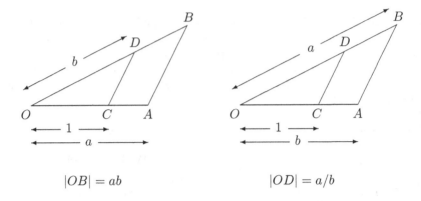

$$|OB| = ab \qquad\qquad |OD| = a/b$$

FIGURE 6.3. Construction of ab and a/b.

In Figure 6.3 the two diagrams are identical, apart from the assignment of the segments that have lengths a, b, and 1. The lines AB and CD are parallel, and thus the triangles OAB and OCD are similar. (The angle between OA and OB is chosen arbitrarily.) It follows from the similarity of triangles OAB and OCD that

$$\frac{|OD|}{|OC|} = \frac{|OB|}{|OA|}. \tag{6.1}$$

In the diagram on the left of Figure 6.3, $|OA| = a$, $|OD| = b$, and $|OC| = 1$, and it then follows from (6.1) that

$$\frac{b}{1} = \frac{|OB|}{a},$$

and hence $|OB| = ab$. In the diagram on the right of Figure 6.3, $|OB| = a$, $|OA| = b$, and $|OC| = 1$, and it then follows from (6.1) that

$$\frac{|OD|}{1} = \frac{a}{b},$$

so that $|OD| = a/b$. ■

Remark 6.1.1 Through the repeated use of Constructions 5 and 6, we can construct any length that can be obtained by beginning with a finite number of given lengths and carrying out a finite number of applications of addition, subtraction, multiplication, and division. ■

Construction 7 *Divide a given line segment into n equal parts.*

Figure 6.4 illustrates the case $n = 3$. We begin with the line AB, and

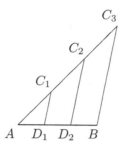

FIGURE 6.4. Divide the line segment AB into three equal parts.

draw a second line through A, making any angle with AB. Beginning at A we mark off n equal segments on the second line, AC_1, C_1C_2, and so on, ending with the segment C_{n-1}, C_n. We join C_n to B. Then, using Construction 4, through each point C_1, C_2, and so on, up to C_{n-1}, we draw a line parallel to the line C_nB. We see that the triangles AC_1D_1, AC_2D_2, and so on, are all similar. Thus in Figure 6.4 we have

$$\frac{|AD_2|}{|AD_1|} = \frac{|AC_2|}{|AC_1|} = 2 \quad \text{and} \quad \frac{|AB|}{|AD_1|} = \frac{|AC_3|}{|AC_1|} = 3.$$

We can obviously extend this argument if $n > 3$. This justifies the construction.   ■

Construction 8 *Bisect a given angle.*

Let the given angle be denoted by angle BAC, where $|BA| = |AC|$. We use the compasses to construct a point D such that $|BD| = |CD|$. (See Figure 6.5.) Then the line AD bisects the angle BAC.

We observe that the triangles DAB and DAC have three corresponding sides equal. Thus they are congruent, and so the angles DAB and DAC are equal.   ■

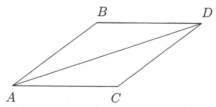

FIGURE 6.5. The line DA bisects the angle BAC.

Construction 9 *Inscribe a circle inside a given triangle.*

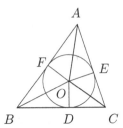

FIGURE 6.6. The incircle for the triangle ABC.

Using Construction 8, draw the lines that bisect angles ABC and ACB, and denote their point of intersection by O. (See Figure 6.6.) From O, draw perpendiculars to each side of the triangle, and let them meet BC, CA, and AB at D, E, and F, respectively. Then O is the center of the inscribed circle and its radius is $|OD|$, which is equal to $|OE|$ and $|OF|$.

The inscribed circle is called the *incircle* of triangle ABC, and O is called the *incenter*. Let us compare the triangles OFB and ODB. The angles OBF and OBD are equal, and the angles OFB and ODB are also equal, both being right angles. Since also the triangles OFB and ODB have the common side OB, the two triangles are congruent, and thus $|OF| = |OD|$. Similarly, we can show that the triangles ODC and OEC are congruent, and deduce that $|OD| = |OE|$. Thus we can draw a circle with center O, and radius $|OD| = |OE| = |OF|$. The sides of the triangle, BC, CA, and AB meet the radii OD, OE, and OF, respectively, at right angles. We say that the sides of the triangle are tangents to the circle. Finally, let us compare the triangles OEA and OFA. They are right-angled triangles with two corresponding sides equal, since the side OA is common to both triangles and $|OE| = |OF|$. Hence, by Pythagoras's theorem (see Section 1.2), we have

$$|AE|^2 = |OA|^2 - |OE|^2 = |OA|^2 - |OF|^2 = |AF|^2.$$

Since $|AE| = |AF|$, the triangles OEA and OFA are congruent, and so OA bisects the angle BAC. Thus the bisectors of the three angles of a triangle are *concurrent*, that is, they meet in a common point. ■

Construction 10 *Draw a circle that passes through three points that do not lie in a straight line.*

Let the three points be denoted by A, B, and C. Using Construction 3, we draw the perpendicular bisectors of AB and BC. These two perpendicular bisectors must intersect at some point, say O. Since $|OA| = |OB| = |OC|$, the points A, B, and C lie on a circle whose center is O. Since O is equidistant from C and A, it lies on the perpendicular bisector of CA, and thus the three perpendicular bisectors are concurrent at O. ■

It is clear from Construction 10 that three points A, B, and C not on a straight line determine a *unique* circle that passes through all three points. This is called the *circumcircle* of the triangle ABC, and its center is called the *circumcenter*. In general, a given fourth point, D, will not lie on the circle that passes through A, B, and C. A quadrilateral $ABCD$ for which, unusually, all four points lie on a circle, is called a *cyclic quadrilateral.*

Construction 11 *Given a line segment of length 1, construct a line segment whose length is the golden ratio,* $\alpha = \frac{1}{2}(\sqrt{5} + 1)$.

Let $|AB| = 1$. We use Construction 1 to create the square $ABCD$, as depicted in Figure 6.7, and use Construction 3 to bisect the line CD at E. Then, by Pythagoras's theorem (see Section 1.2) we have

$$|EB|^2 = |BC|^2 + |CE|^2 = 1 + \frac{1}{4} = \frac{5}{4},$$

and hence $|EB| = \frac{1}{2}\sqrt{5}$. We now use the compasses, centered on E, to draw a circle of radius $|EB|$ to cut the extended line DC at F. Then

$$|DF| = |DE| + |EF| = |DE| + |EB| = \frac{1}{2} + \frac{1}{2}\sqrt{5} = \alpha. \qquad \blacksquare$$

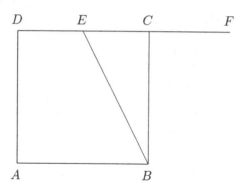

FIGURE 6.7. Construction of the golden ratio $|DF|/|DC|$.

Of all the regular polygons, the equilateral triangle and the regular hexagon are the easiest to construct.

Construction 12 *Construct a regular hexagon in a circle of radius 1.*

We choose a point A_1 on the circumference of the circle, and use the compasses with radius 1 and center A_1 to draw an arc that cuts the circle at a point A_2. See Figure 6.8, where O marks the center of the circle. We next center the compasses on A_2 and, with the same radius, draw an arc

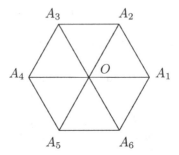

FIGURE 6.8. Construction of the regular hexagon.

that cuts the circumference at a point A_3, as shown in Figure 6.8. We repeat this process three more times to complete the construction. We note that the figure contains six congruent equilateral triangles, namely OA_1A_2, OA_2A_3, and four others. If we pick out every second point of the hexagon $A_1A_2A_3A_4A_5A_6$, we obtain an equilateral triangle.   ■

Construction 13 *Construct a regular decagon in a circle of radius 1.*

Let O be the center of a circle of radius 1, let $|OA| = |OB| = 1$ and let AB be a chord of a regular decagon, the regular polygon with ten sides. Thus the angle AOB is $\frac{\pi}{5}$. See Figure 6.9. The two remaining angles in triangle OAB are equal, and must both be $\frac{2\pi}{5}$, since the sum of the angles in a triangle is equal to 2π. The point C is chosen on OA so that $|BA| = |BC|$. Therefore, the two angles BCA and BAC are both $\frac{2\pi}{5}$, and consequently angle CBA is $\frac{\pi}{5}$. This means that angle CBO is also $\frac{\pi}{5}$, and so $|CB| = |CO|$. Thus the triangles AOB and ABC are both similar to the triangle FAG in Figure 3.7, which depicts the pentagram. On comparing the two similar triangles in Figure 6.9, we have

$$\frac{|AB|}{|AC|} = \frac{x}{y} = \frac{x}{1-x} = \frac{|OA|}{|AB|} = \frac{1}{x}. \tag{6.2}$$

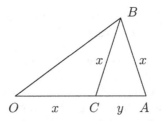

FIGURE 6.9. Construction of the regular decagon.

As we found in Section 3.3 (see (3.39)),

$$\frac{x}{y} = \alpha = \frac{1}{2}\left(\sqrt{5}+1\right),$$

the golden ratio. Since $y = 1 - x$, we find that

$$x = \frac{\alpha}{\alpha+1} = \frac{\alpha^2-1}{\alpha+1} = \alpha - 1.$$

We can obtain the golden ratio α by ruler and compasses using Construction 11, and thus we can construct $x = \alpha - 1$, the length of the side of a regular decagon in a circle of radius 1. The construction of the decagon is then easily completed. We can obtain the regular pentagon by connecting every second point of the decagon, and can also construct the star of Pythagoras, which we discussed in Section 3.3.   ■

Given two positive numbers a and b, the number

$$A(a,b) = \frac{1}{2}(a+b) \tag{6.3}$$

is called their *arithmetic mean*. If $0 < b \le a$, we have

$$2b \le a + b \le 2a.$$

Thus

$$b \le \frac{1}{2}(a+b) \le a, \tag{6.4}$$

so that $A(a,b)$ lies between a and b. We define the *geometric mean* of a and b as

$$G(a,b) = \sqrt{ab}. \tag{6.5}$$

If $0 < b \le a$ we can show that $0 < b^2 \le ab \le a^2$, and thus

$$0 < b \le \sqrt{ab} \le a, \tag{6.6}$$

so that $G(a,b)$ lies between a and b. If we compute the arithmetic and geometric means of several pairs of positive numbers a and b, we always find that $A(a,b) \ge G(a,b)$. To verify this, consider

$$\frac{1}{2}(a+b) - \sqrt{ab} = \frac{1}{2}\left(a - 2\sqrt{ab} + b\right) = \frac{1}{2}\left(\sqrt{a} - \sqrt{b}\right)^2 \ge 0. \tag{6.7}$$

Let $x = \sqrt{ab}$, so that $ab = x^2$. On dividing throughout by bx, we obtain

$$\frac{a}{x} = \frac{x}{b}, \tag{6.8}$$

and we say that the geometric mean x is the *mean proportional* of a and b. I will mention only one more mean, although the list of means is endless. This is the *harmonic mean* of a and b, defined by

$$H(a,b) = \frac{2ab}{a+b}. \tag{6.9}$$

The arithmetic, geometric, and harmonic means have been studied since at least the time of Pythagoras, in the sixth century BC. If we divide both the numerator and the denominator of the fraction on the right of (6.9) by ab, we obtain

$$H(a,b) = 2\Big/ \left(\frac{1}{a} + \frac{1}{b}\right).$$

Thus the harmonic mean is the reciprocal of the arithmetic mean of $1/a$ and $1/b$, that is

$$H(a,b) = 1\Big/ A\left(\frac{1}{a}, \frac{1}{b}\right). \tag{6.10}$$

It is also not hard to verify that

$$\frac{H(a,b)}{G(a,b)} = \frac{G(a,b)}{A(a,b)}. \tag{6.11}$$

Thus

$$G(a,b) = \Big(A(a,b)H(a,b)\Big)^{1/2},$$

so that the geometric mean of a and b is itself the geometric mean of the arithmetic and harmonic means of a and b.

Construction 14 *Construct the arithmetic and geometric means of two line segments AB and BC.*

In Book III of his *Mathematical Collection*, Pappus of Alexandria (third century AD) gives the construction that is shown in Figure 6.10. This depicts the case $|AB| > |BC|$. The arithmetic mean is easily found by using Construction 3 to find O, the midpoint of AC. Then both $|AO|$ and $|OC|$ give the required arithmetic mean of $|AB|$ and $|BC|$. Next, using Construction 1, we draw a perpendicular to AC, passing through B, to meet the semicircle with diameter AC at the point D. Then $|BD|$ is the geometric mean of $|AB|$ and $|BC|$. The proof relies on similar right-angled triangles, and we first show that the angle ADC is a right angle. We argue that in triangle OAD, $|OA| = |OD|$, both being radii of the semicircle that passes through A, D, and C. Thus the angles OAD and ODA are equal, and we will denote them by α. Similarly, in triangle OCD, since OC and OD are radii of the semicircle, the angles OCD and ODC are equal, and we will denote them by β. Since the sum of the angles of triangle ADC, which equals two right angles, is also equal to $2\alpha + 2\beta$, it follows that $\alpha + \beta$ is a right angle. Thus angle ADC is a right angle. It follows that the triangles ADC, ABD, and DBC, having corresponding angles equal, are similar right-angled triangles. We deduce from the similarity of the two smaller triangles that

$$\frac{|AB|}{|BD|} = \frac{|BD|}{|BC|}, \qquad (6.12)$$

so that $|BD|$, being the mean proportional of $|AB|$ and $|BC|$, is their geometric mean.   ∎

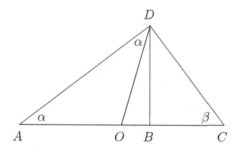

FIGURE 6.10. $|OD|$ is the arithmetic mean of $|AB|$ and $|BC|$, and $|BD|$ is their geometric mean.

Construction 15 *Construct the harmonic mean of two line segments AB and BC.*

This construction is also given by Pappus in Book III of his *Mathematical Collection*. We amend Figure 6.10 by drawing a line from B, perpendicular to OD, meeting OD at the point E, as in Figure 6.11. Then $|ED|$ is the harmonic mean of $|AB|$ and $|BC|$. To justify this, we observe that the triangles DEB and DBO are similar, and thus

$$\frac{|ED|}{|DB|} = \frac{|DB|}{|OD|}.$$

This shows that $|DB|$, which is the geometric mean of $|AB|$ and $|BC|$, is a mean proportional to $|ED|$ and $|OD|$, and since $|OD|$ is the arithmetic mean of $|AB|$ and $|BC|$, it follows from (6.11) that $|ED|$ is the harmonic mean of $|AB|$ and $|BC|$.   ∎

We observe from Figure 6.11 that

$$|OD| > |BD| > |ED|,$$

showing an ordering of the arithmetic, geometric, and harmonic means. This may be justified algebraically (see Problem 6.1.7).

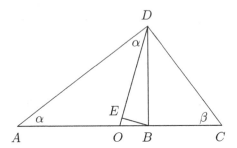

FIGURE 6.11. $|ED|$ is the harmonic mean of $|AB|$ and $|BC|$.

Problem 6.1.1 To construct a regular octagon, draw a circle with center O, and inscribe a square $ABCD$ in it, giving four vertices of the octagon. (One way is to draw two diameters of the circle that are mutually perpendicular.) Use Construction 3 to find E, the midpoint of AB, and show that the point where OE cuts the circle gives a fifth vertex of the octagon. Finally, construct the three remaining vertices of the octagon.

Problem 6.1.2 Construct an equilateral triangle and a regular pentagon that have one vertex in common within a circle of radius 1. (See Constructions 12 and 13.) Hence construct a regular polygon with 15 sides.

Problem 6.1.3 If $0 < b \le a$, verify that $0 < b^2 \le ab \le a^2$ and hence show that

$$b \le \sqrt{ab} \le a.$$

Problem 6.1.4 Show that for the arithmetic mean,

$$A(\lambda a, \lambda b) = \lambda A(a, b),$$

where λ, a, and b are any positive numbers. Any mean satisfying this property is called *homogeneous*. Show that the geometric and harmonic means are also homogeneous.

Problem 6.1.5 In Figure 6.11, in which we have $|AB| > |BC|$, prove that $|ED| > |BC|$.

Problem 6.1.6 Deduce from (6.9) that if $0 < b \le a$, then

$$b \le H(a, b) \le a.$$

Problem 6.1.7 Verify that if $0 < b \le a$,

$$b \le H(a, b) \le G(a, b) \le A(a, b) \le a,$$

where A, G, and H denote the arithmetic, geometric, and harmonic means.

6.2   Unsolvable Problems

In Section 6.1 we saw how to construct regular n-sided polygons for $n = 3$, 4, 5, 6, 8, and 10, that is, the equilateral triangle, the square, and the regular pentagon, hexagon, octagon, and decagon. However, neither the regular *heptagon*, which has seven sides, nor the regular *nonagon*, which has nine sides, can be constructed using ruler and compasses. Heron of Alexandria found an approximate construction for the heptagon, and the famous painter, engraver, *and* mathematician Albrecht Dürer (1471–1528) found an approximate construction for the nonagon. (See Eves [9].) It may seem surprising that the regular nonagon cannot be constructed by ruler and compasses. We can begin by constructing an equilateral triangle, say ABC, in a circle with center O. Then we could complete the construction by trisecting the angles AOB, BOC, and COA. Alas, we cannot do that! For apart from some special cases, including a right angle, we cannot trisect an angle using ruler and compasses. The trisection of an angle is one of the three famous unsolved problems of Greek mathematics.

In Book IV of his *Elements*, Euclid described the construction of regular polygons of 3, 4, 5, 6, and 15 sides. We can extend these constructions by carrying out repeated bisections, using the same device as we employed in Problem 6.1.1 to construct the regular octagon from the square. Thus we can construct regular polygons with 2^n sides for any $n \geq 2$, or $m \times 2^n$ sides, where $m = 3$, 5, or 15, and n is any nonnegative integer. Since the time of Euclid, the only substantial addition to our knowledge of the ruler and compass construction of regular polygons was made by C. F. Gauss, who had the last word on this topic. Gauss showed that the construction of a regular polygon with a prime number of sides p is possible if and only if p is a prime number of the form $f_n = 2^{2^n} + 1$, for $n \geq 0$. These are the Fermat numbers, defined in (4.22), and we obtain the prime values 3, 5, 17, 257, and 65,537, corresponding to $n = 0$, 1, 2, 3, and 4. As was stated in Section 4.3, no other Fermat primes are known. (A star-like figure based on the regular polygon with 17 sides is inscribed on the plinth of Gauss's statue in Braunschweig, the city of his birth.)

The second of the three famous unsolved problems of Greek mathematics is the duplication of the cube. It is easy to construct a square whose area is twice that of a given square. For if we have a square of side a, its area is a^2, and we see from Pythagoras's theorem that its diagonal is of length $\sqrt{2}a$. Using ruler and compasses, we can construct a square on this diagonal whose area is $2a^2$. Analogously, given a cube of side a, can we construct a number b such that $b^3 = 2a^3$? Hippocrates of Chios showed in the fifth century BC that this is equivalent to finding numbers b and c such that

$$\frac{2a}{c} = \frac{c}{b} = \frac{b}{a}. \tag{6.13}$$

For on multiplying the first equality in (6.13) by bc, and the second equality by ab, we find that (6.13) is equivalent to

$$2ab = c^2 \quad \text{and} \quad ac = b^2. \tag{6.14}$$

The second equation in (6.14) is equivalent to

$$a^2c^2 = b^4. \tag{6.15}$$

On multiplying the first equation in (6.14) by a^2, and using (6.15), we obtain

$$2a^3b = b^4,$$

which indeed reduces to

$$b^3 = 2a^3. \tag{6.16}$$

The numbers b and c sought by Hippocrates, as in (6.13), are said to be mean proportionals to the numbers a and $2a$. Although the construction of *one* mean proportional to two numbers (their geometric mean) is easy, the construction of two mean proportionals defied the considerable ingenuity of generations of Greek mathematicians, and surely this included some of the cleverest people who have ever existed.

The last of the three classical unsolved problems of ancient Greek mathematics is called the squaring of the circle. By their construction of the geometric mean, the Greeks had shown how to construct a square whose area is equal to the area of a given rectangle. They also tried, in vain, to construct a square whose area is the same as that of a given circle. You may think that this seems too ambitious, since unlike the perimeter of a rectangle, the perimeter of a circle is a curve, and so its area is much more difficult to reconcile with that of a square. However, the greatest of the Greek mathematicians, Archimedes of Syracuse (287–212 BC), showed that the area of a segment of a parabola can be expressed as a rational multiple of the area of a certain triangle. Since the the circle appears to be a simpler object than the parabola, it is understandable if this encouraged the belief that the squaring of the circle was achievable.

Let us now give a brief account of *conic sections*. We begin with a straight line ℓ that intersects a given vertical line at a point O. (See Figure 6.12.) A *right circular cone* is the three-dimensional figure that is created by rotating ℓ around the vertical line. Both ℓ and the vertical line are infinite in length, and thus the cone extends to infinity both above and below the central point O, which is called the *vertex* of the cone. In everyday language, the word "cone" is commonly used to denote just one half of the figure we have just described. The vertical line in Figure 6.12 is called the *axis* of the cone. Every line that passes through O and lies on the surface of the cone, like the line ℓ, is called a *generator*. The cone is determined uniquely by the angle that ℓ makes with the axis. This angle, which we will take to be less than a right angle, is denoted by α in Figure 6.12.

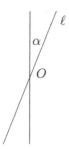

FIGURE 6.12. A right circular cone is generated by rotating the straight line ℓ around the vertical line.

Conic sections, as the name suggests, are the curves that are obtained from sections of the cone by planes. Consider a plane that cuts the axis of the cone at an angle $\beta \leq \frac{\pi}{2}$. The simplest and not so interesting case is that of the plane cutting through the vertex. In this case, if $\beta < \alpha$, the plane cuts the cone in a pair of generators, and if $\beta = \alpha$, the plane cuts the cone in a single generator. If $\beta > \alpha$, the plane obviously cuts the cone only at O. Now let us consider the curves that are obtained when a plane forming an angle β with the axis does not pass through the vertex O. If $\beta < \alpha$, the plane cuts both halves of the cone and we obtain a *hyperbola*, which thus has two branches. If $\beta = \alpha$, the plane is parallel to a generator of the cone. It cuts only one half of the cone and the resulting curve of intersection is the *parabola*. If $\beta > \alpha$, the plane cuts the cone through one half only, and we obtain an *ellipse*. In particular, if $\beta = \frac{\pi}{2}$ we obtain the *circle* as a special case of an ellipse.

Archimedes used a most ingenious construction to prove that the area of a segment of a parabola is equal to $\frac{4}{3}$ times the area of a triangle whose base is the same as the length of the parabolic segment and that has the same height. Figure 6.13 shows the parabolic segment and the triangle ABC that has the same base and height as the segment. Archimedes constructed two further triangles, shown as triangles ADB and BEC in Figure 6.13. The point D is chosen so that the triangle ADB has the same height as the parabolic segment with base AB. Similarly, triangle BEC, with base BC, has the same height as the parabolic segment with base BC. Archimedes proved that the sum of the areas of triangles ADB and BEC is one quarter of the area of triangle ABC, and a fine account of this proof is given in Edwards [7]. Archimedes continued this process, constructing four triangles with bases AD, DB, BE, and EC, whose combined area is one-quarter of the combined areas of triangles ADB and BEC, and so is $\frac{1}{16}$ of the area of triangle ABC. Archimedes thought of this construction being continued indefinitely, and proved that the area of the original parabolic segment that is not covered by one of his triangles tends to zero. He deduced that

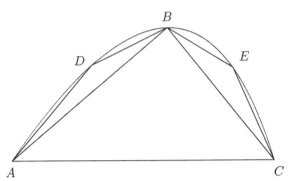

FIGURE 6.13. The area of the parabolic segment is $\frac{4}{3}$ times the area of the triangle ABC.

the area of the parabolic segment ABC is S times the area of the triangle ABC, where

$$S = 1 + \frac{1}{4} + \frac{1}{4^2} + \frac{1}{4^3} + \cdots. \tag{6.17}$$

This is an infinite geometric series, and we see from (4.19) that $S = \frac{4}{3}$, thus verifying this wonderful result of Archimedes.

By the nineteenth century, some areas of mathematics had been developed that were not known to the ancient Greek mathematicians. It was one of these, the theory of equations, that led to proofs that the three famous long-standing problems of Greek mathematics are insoluble. Consider the polynomial equation

$$a_0 z^n + a_1 z^{n-1} + \cdots + a_{n-1} z + a_n = 0, \tag{6.18}$$

where the coefficients $a_0, a_1, \ldots, a_n$ are all real and a_0 is nonzero. This equation is easily solved if $n = 1$, when it is called a linear equation. When $n = 2$ we have a quadratic equation, whose complete solution is discussed in Section 1.5. Some quadratic equations were solved by Babylonian mathematicians as early as the second millennium BC. When $n = 3$, 4, and 5 we have a cubic, quartic, and quintic equation, respectively. It was not until the sixteenth century AD that the general cubic and quartic equations were solved. When I say "solved," I mean that solutions of the equation can be obtained by carrying out a finite number of arithmetical operations, specifically additions, subtractions, multiplications, divisions, and the extraction of square roots, cube roots, and so on, beginning with the coefficients a_0, $a_1, \ldots, a_n$. By the early eighteenth century, P. Ruffini (1765–1822) and N. H. Abel (1802–1829) independently proved that the general quintic cannot be solved, in the sense defined above. Although it may seem rather negative to show that something cannot be done, this was a very important achievement in the history of mathematics.

Recall from Definition 2.5.2 that a complex number that is a solution of an equation of the form (6.18) where the coefficients are all integers is called algebraic, and a complex number that is not a solution of such an equation is called transcendental. The breakthrough in the long quest to settle the three most famous unsolved ruler and compass problems came with the establishing of the following two key results in the theory of equations, which are stated in Eves [9].

Theorem 6.2.1 Beginning with a line segment of unit length, the length of any line segment constructed by any sequence of operations using ruler and compasses is an algebraic number.   ■

Theorem 6.2.2 Beginning with a line segment of unit length, it is impossible to construct by any sequence of operations using ruler and compasses a line segment whose length is a solution of a cubic equation that has integer coefficients but has no rational solution.   ■

As we will see, Theorem 6.2.2 settles the question of the duplication of the cube in the negative. In (6.16) let us write $a = 1$ and $b = x$. Then we need to construct a number x such that $x^3 - 2 = 0$. If there were a rational solution of this equation, we could write it as

$$x = \frac{p}{q}, \quad \text{where} \quad \frac{p^3}{q^3} = 2,$$

where p and q are positive integers, and we can assume that p and q have no common factor greater than 1. Thus

$$p^3 = 2q^3, \tag{6.19}$$

and so p must be even. Let us write $p = 2p_1$ in (6.19), so that

$$8p_1^3 = 2q^3,$$

and thus

$$4p_1^3 = q^3,$$

showing that q must be even. Notice that, beginning with the assumption that the equation $x^3 - 2 = 0$ has a solution of the form $x = p/q$, where p and q have no common factor greater than 1, we have shown that p and q have the common factor 2. This shows that our assumption is untenable, and that this cubic equation with integer coefficients does not have a rational solution. It follows from Theorem 6.2.2 that the duplication of the cube cannot be achieved by a ruler and compass construction.

Theorem 6.2.2 can also be used to show that not every angle can be trisected by using a ruler and compass construction. One needs only a little knowledge of trigonometry to follow a proof of this. See Eves [9].

The area of a circle of radius r is πr^2. Thus the area of a circle of unit radius is π. To "square the circle," we need to construct a square with side of length $\sqrt{\pi}$. In 1882, C. L. F. Lindemann (1852–1939) proved that π is transcendental, following up the work of C. Hermite (1822–1901), who showed in 1873 that e is transcendental. Thus, by Theorem 6.2.1, π cannot be constructed by a ruler and compass construction. Now if $\sqrt{\pi}$ were constructible by ruler and compasses, we could redraw Figure 6.10, beginning with $|BD| = \sqrt{\pi}$ and $|BC| = 1$. Then, since $|BD|$ is the geometric mean of $|AB|$ and $|BC|$, it follows that $|AB| = \pi$. Thus if $\sqrt{\pi}$ could be constructed by ruler and compasses, so could π, which is not constructible. This shows the impossibility of squaring the circle by using a ruler and compass construction.

Problem 6.2.1 Verify from (4.19) that

$$\frac{1}{3} = \frac{1}{4} + \frac{1}{4^2} + \frac{1}{4^3} + \cdots,$$

and hence show how we can get as close as we wish to trisecting an angle using ruler and compasses by using repeated bisections of an angle. Note that this does *not* show that we can trisect an angle using a *finite* number of ruler and compass operations.

6.3   Properties of the Triangle

Consider three line segments whose lengths are a, b, and c, where $a \geq b \geq c$. It is obvious that these three segments can be fitted together to form a triangle if and only if $b + c > a$, and if a triangle can be formed, it must be unique. Thus if we have two triangles whose sides are a, b, and c, they must be congruent. Suppose we have two triangles that both have sides b and c, and have the same angle between these sides. We say that the two triangles have two sides and the included angle equal. Then we can fit one triangle on top of the other, and it is clear that the third sides must be equal. Therefore, having two sides and the included angle equal is a second condition that gives congruent triangles. There is a third congruence condition, two triangles being congruent if they have the same angles and have one corresponding side equal. For again we can verify that they are congruent by fitting one triangle on top of the other. In practice, we need only verify that two pairs of angles are equal, since equality of the third angles follows from the result we discuss in the next paragraph.

One of the most basic properties of a triangle is that the sum of its three angles is 2π, which is equal to two right angles. A proof of this is given in Problem 1.2.1. Another proof, which is equivalent to this, can be realized by cutting out a paper copy of the triangle and folding it, as shown in Figure

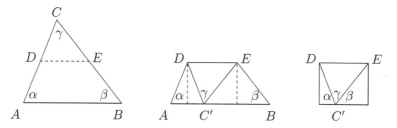

FIGURE 6.14. The sum of the angles in a triangle equals 2π.

6.14. We fold the triangle along a line that bisects two of the sides (see the dotted line DE in the left-hand diagram in Figure 6.14), so that the vertex denoted by C meets the side AB at C'. The line DE is parallel to AB, since the triangles CDE and CAB are similar, and hence the angles CDE and CAB are equal. In the middle diagram we see that the triangle DAC' is isosceles, since $|DA| = |DC'|$, and so the angles DAC' and $DC'A$ are equal. Similarly, the angles $EC'B$ and EBC' are equal. We next fold the triangle along the two vertical dotted lines shown in the middle diagram in Figure 6.14, so that the vertices A and B both coincide with C'. Then we see in the right-hand diagram in Figure 6.14 that the three angles of the original triangle ABC add up to 2π, or two right angles.

We proved in Construction 10 that the perpendicular bisectors of the three sides of a triangle are concurrent. The point of concurrence, which we will denote by A, is called the circumcenter. It is the center of a circle, the circumcircle, that passes through the vertices of the triangle.

In Construction 9 we showed that the bisectors of the three angles of a triangle are concurrent. This point of concurrence, which we will denote by O, is called the *incenter*. It is the center of a circle, the incircle, that is inscribed in the triangle, touching all three sides.

There are two other very well known points of concurrence in a triangle, which I will justify in the next section. One is the orthocenter, which we will denote by B, where the three perpendiculars from the vertices of the triangle to the opposite sides meet. (These perpendiculars are also called the *altitudes* of the triangle.) The other is the *centroid*, which we will denote by C, where the three *medians* of the triangle intersect. A median is a line joining a vertex of the triangle to the midpoint of the opposite side. The points A, B, and C are displayed in Figure 6.15. In this figure, A_1, A_2, and A_3 denote the vertices of the triangle, B_1, B_2, and B_3 denote the points where the perpendiculars from the vertices of the triangle meet the opposite sides, and C_1, C_2, and C_3 denote the midpoints of the sides of the triangle. The figure displays the three lines that are concurrent at B, and those that are concurrent at C are shown as dotted lines. The three lines that are concurrent at A are omitted from the figure for the sake of clarity,

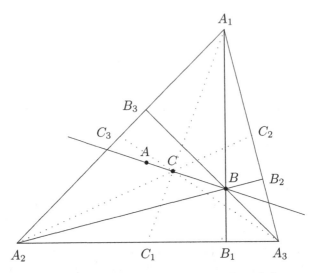

FIGURE 6.15. The circumcenter A, the orthocenter B, and the centroid C all lie on the Euler line.

and especially to help us see that the three points A, B, and C all lie on the same line. The line that contains A, B, and C is called the Euler line, after Leonhard Euler. The centroid trisects each median, meaning that

$$\frac{|CC_1|}{|A_1C_1|} = \frac{|CC_2|}{|A_2C_2|} = \frac{|CC_3|}{|A_3C_3|} = \frac{1}{3}.$$

Amazingly, the centroid also trisects AB, for we have

$$\frac{|AC|}{|AB|} = \frac{1}{3}.$$

We will verify both of these trisection properties of the centroid in the next section.

In Construction 9 we saw how given a triangle ABC we can draw the incircle, which touches BC, CA, and AB on the inside of the triangle. By making a simple modification of this process, we can construct three *excircles* that touch the line segments BC, CA, and AB, or their extensions, *outside* the triangle. Figure 6.16 shows one of these excircles. We bisect the angles DAC and ECA, and let the bisectors intersect at the point O. Then, using the same argument as we used in Construction 9, we see that O is equidistant from the lines BD, BE, and AC. Thus OB is the bisector of the angle DBE.

As we have already seen, there are many interesting and surprising properties possessed by all triangles, and some are more difficult to justify than others. It is not surprising that many of these properties have been known

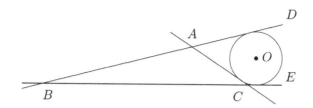

FIGURE 6.16. One of the three excircles of triangle ABC.

since the early part of the millennium of ancient Greek mathematics, and it is easy to imagine how they would have been discovered and rediscovered many times in an age when ruler and compasses were everyday tools of mathematicians. If we draw several arbitrarily chosen triangles and find every time that, for example, the perpendiculars from the vertices of the triangle to the opposite sides are concurrent, it is natural to believe that the result must hold for every triangle. Then we cannot rest until we find a proof!

One most unusual property of the triangle is named after Napoleon Bonaparte (1769–1821). Since Napoleon was a keen amateur geometer, some historians of mathematics believe that he did indeed discover the result, although at least some of its ingredients were known long before his time. We begin with any triangle ABC and draw an equilateral triangle on each of its sides, on the outside of the triangle, as in Figure 6.17. Then the following statements all hold.

1. The centroids of the three equilateral triangles are themselves the vertices of an equilateral triangle.

2. The three lines AA_1, BB_1, and CC_1 are collinear, meeting at a point P, called the *isogonic center* of the triangle ABC.

3. The three line segments AA_1, BB_1, and CC_1 are of equal length, and

$$|AA_1| = |BB_1| = |CC_1| = |PA| + |PB| + |PC|. \qquad (6.20)$$

4. The angles B_1PC_1, C_1PA_1, and A_1PB_1 are all equal, and thus each has the value $\frac{2\pi}{3}$.

5. The circumcircles of the three equilateral triangles all pass through the isogonic center P.

In response to a challenge by Pierre de Fermat, E. Torricelli (1608–1647) found the isogonic center as the solution to the problem of finding a point in the plane of a triangle that minimizes the sum of its distances from the three vertices. Obviously Torricelli's result was obtained several generations

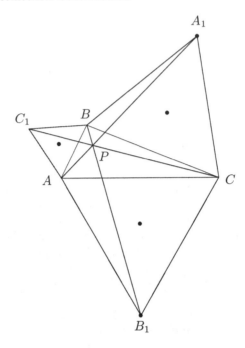

FIGURE 6.17. Napoleon's theorem.

before Napoleon's time. Eves [9] states that the isogonic center was the first notable point of the triangle to be discovered since the era of ancient Greek mathematics. Figure 6.17 gives a ruler and compasses construction of the isogonic center.

In the early nineteenth century, a further set of properties of the triangle was obtained by Karl Feuerbach (1800–1834). This is embodied in the *nine-point circle* theorem, which P. J. Davis [5] says goes back, in part, to J. V. Poncelet (1788–1867) in 1820. I now state this theorem and will omit the proof.

Theorem 6.3.1 The following nine points, related to the triangle with vertices A_1, A_2, and A_3, all lie on the same circle, and are shown in Figure 6.18:

1. The points B_1, B_2, and B_3 where the three altitudes from the vertices of the triangle meet the opposite sides.

2. The midpoints of the three sides of the triangle, C_1, C_2, and C_3.

3. The points D_1, D_2, and D_3, the midpoints along the altitudes from the vertices of the triangle to the orthocenter B.   ∎

There are other notable points that are on this circle. For example, the nine-point circle touches the incircle of the triangle $A_1 A_2 A_3$, say at E_0,

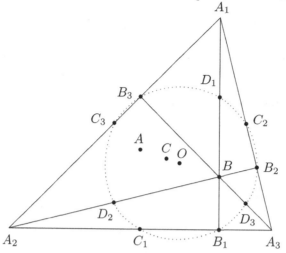

FIGURE 6.18. The nine-point circle.

and all three excircles of the triangle, say at E_1, E_2, and E_3, and thus the four points E_0, E_1, E_2, and E_3 also lie on the nine-point circle.

Let O denote the midpoint of the line joining A, the circumcenter of the triangle, and B, the orthocenter of the triangle. Then, in Figure 6.18, the triangles OBD_1 and ABA_1 are similar, since they have the common angle ABA_1 and

$$\frac{|OB|}{|AB|} = \frac{|D_1B|}{|A_1B|} = \frac{1}{2}.$$

It follows from the similarity of the triangles that

$$\frac{|D_1O|}{|A_1A|} = \frac{1}{2}. \tag{6.21}$$

Since $|A_1A| = |A_2A| = |A_3A|$, we see that $|D_1O| = |D_2O| = |D_3O|$, showing that O is the center of the circle that passes through D_1, D_2, and D_3. This defines the nine-point circle, and it follows from (6.21) that its radius is half that of the circumcircle of triangle $A_1A_2A_3$. Note that O, the center of the nine-point circle, lies on the Euler line.

Problem 6.3.1 Draw Figure 6.17 for the special case in which the triangle ABC is itself an equilateral triangle.

Problem 6.3.2 In Figure 6.17 the equilateral triangles are drawn on the sides of triangle ABC, on the outside of the triangle. Investigate what happens if you draw the same three triangles on the other sides of BC, CA, and AB.

Problem 6.3.3 Draw any quadrilateral, and draw a square on each of its sides, on the outside of the quadrilateral. Finally, draw the two line segments obtained by joining the centers of the squares on opposite sides of the quadrilateral. What can you observe about these two line segments?

Problem 6.3.4 Consider Problem 6.3.3 again, for the special case in which the quadrilateral is a parallelogram.

6.4   Coordinate Geometry

In the seventeenth century geometry was reborn, with the introduction of the new and powerful methods of *coordinate geometry*. These methods are credited to René Descartes (1596–1650) and his contemporary Pierre de Fermat, although the underlying ideas have earlier origins. Descartes realized that every point in the plane can be uniquely defined by its distance plus direction from each of two axes set at right angles. Figure 6.19 shows the horizontal x-axis and the vertical y-axis, marked off at integer points. Each point in the plane of the diagram has an ordered pair of x and y coordinates. These are called Cartesian coordinates, named after Descartes, to distinguish them from other coordinate systems.

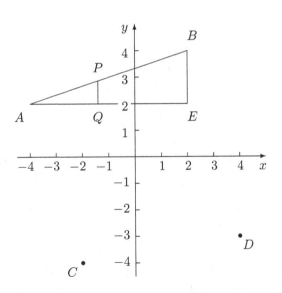

FIGURE 6.19. Cartesian coordinates.

In Figure 6.19, A has coordinates $(-4, 2)$, B has coordinates $(2, 4)$, and the points C and D have coordinates $(-2, -4)$ and $(4, -3)$, respectively.

In triangle ABE, the lines AE and BE are parallel to the x- and y-axes, respectively. Thus E is the point $(2,2)$. Let P be the point on the straight line AB with coordinates (x, y). The point Q is vertically below P, and is on the line AE. Therefore, Q has coordinates $(x, 2)$. There must be a relation between the two coordinates of the point P, and we will now determine what it is. Since the triangles APQ and ABE are similar, we have

$$\frac{|PQ|}{|AQ|} = \frac{|BE|}{|AE|},$$

which gives

$$\frac{y-2}{x-(-4)} = \frac{4-2}{2-(-4)} = \frac{2}{6} = \frac{1}{3}.$$

On multiplying the latter equation throughout by $3(x+4)$, we obtain

$$3(y-2) = x+4,$$

which we can write in the form

$$x - 3y + 10 = 0. \tag{6.22}$$

Thus the coordinates (x, y) of all points that lie on the line AB satisfy (6.22), which is called the equation of the line AB. We can verify that the coordinates of A and B, namely $(-4, 2)$ and $(2, 4)$, both satisfy (6.22). This gives a reassuring check on our calculations. Every straight line has an equation of the form

$$ax + by + c = 0. \tag{6.23}$$

In particular, the x-axis has the equation $y = 0$, and the y-axis has the equation $x = 0$. Lines of the form $by + c = 0$ are parallel to the x-axis, and lines of the form $ax + c = 0$ are parallel to the y-axis. Apart from lines that are parallel to the y-axis, where $b = 0$ in (6.23), we can divide by b to recast (6.23) in the form

$$y = mx + d, \tag{6.24}$$

where $m = -a/b$ and $d = -c/b$. In (6.24) the constant m is called the *gradient* or *slope* of the straight line. It corresponds to the ratio $|BE|/|AE|$ for the line AB. If in Figure 6.19 we let $A = (x_1, y_1)$ and $B = (x_2, y_2)$, we find that the gradient of AB is

$$m = \frac{y_2 - y_1}{x_2 - x_1}. \tag{6.25}$$

Note that the value of m is unchanged if we interchange (x_1, y_1) and (x_2, y_2), as we should expect. If the gradient is positive, the line slopes upwards from left to right, and if the gradient is negative, the line slopes downwards from left to right. The greater the magnitude of the gradient, the steeper the slope. The number d in (6.24) is called the *intercept*. It marks the point on the y-axis where it is cut by the straight line.

In Figure 6.10 the lines AD and DC meet at right angles. The gradient of the line AD is the positive number $|DB|/|AB|$, and the gradient of DC is the negative number $-|DB|/|BC|$. Thus the *product* of these gradients is equal to -1, since we saw from the similar triangles ABD and DBC (see (6.12)) that $|DB|^2$ is equal to the product of $|AB|$ and $|BC|$. We deduce that, excluding lines that are parallel to the axes, the product of the gradients of any two lines that are perpendicular is equal to -1. For example, as we see from (6.25), the line AC in Figure 6.19 has gradient -3 and is therefore perpendicular to the line AB, which has gradient $1/3$.

Example 6.4.1 Let P and Q have coordinates (x_1, y_1) and (x_2, y_2), and let R have coordinates (x_2, y_1). With this choice of the points P, Q, and R, we see that the angle PRQ is a right angle. (Draw a diagram.) It then follows from Pythagoras's theorem that $|PQ|^2 = |PR|^2 + |QR|^2$, and thus

$$|PQ|^2 = (x_1 - x_2)^2 + (y_1 - y_2)^2. \tag{6.26}$$

On taking the square root we obtain a simple expression for the *distance* $|PQ|$ in terms of the Cartesian coordinates of P and Q.   ∎

Example 6.4.2 Let $ax + by + c = 0$ denote the straight line that joins the points P_1 and P_2, with coordinates (x_1, y_1) and (x_2, y_2), respectively. By this we mean not just the finite line segment that lies between P_1 and P_2, but the line that connects the points and extends indefinitely in both directions. Then we have

$$ax_1 + by_1 + c = 0 \tag{6.27}$$

and

$$ax_2 + by_2 + c = 0. \tag{6.28}$$

If we multiply (6.27) by λ, (6.28) by $1 - \lambda$, and add the two resulting equations, we obtain

$$a(\lambda x_1 + (1-\lambda)x_2) + b(\lambda y_1 + (1-\lambda)y_2) + c = 0.$$

This last equation shows that every point (x, y), where

$$x = \lambda x_1 + (1-\lambda)x_2, \qquad y = \lambda y_1 + (1-\lambda)y_2,$$

lies on the straight line $ax + by + c = 0$. The values of λ between 0 and 1 give the points that lie between P_1 and P_2. In particular, the value $\lambda = \frac{1}{2}$ gives the midpoint of $P_1 P_2$. Can you prove this?   ∎

Example 6.4.3 Let us find the distance between two parallel lines l_1 and l_2 whose equations are

$$ax + by + c_1 = 0 \quad \text{and} \quad ax + by + c_2 = 0, \tag{6.29}$$

where $c_1 \neq c_2$. One way of solving this problem is to begin by choosing any point, say, P_1, on the line l_1. Next, we construct the line l_3 that passes through P_1 and is perpendicular to l_1. Let the lines l_3 and l_2 intersect at the point Q_1. Then $|P_1 Q_1|$ is the distance between l_1 and l_2.

If $a \neq 0$, we can put $y = 0$ in the equation for l_1 in (6.29), and thus choose

$$P_1 = \left(-\frac{c_1}{a}, 0\right). \tag{6.30}$$

Then the line l_3 has gradient b/a, contains P_1, and has equation

$$-bx + ay - \frac{bc_1}{a} = 0. \tag{6.31}$$

On solving equation (6.31) and the second equation in (6.29) simultaneously, we find that Q_1, the point of intersection of l_3 and l_2, is given by

$$Q_1 = \left(\frac{-c_2 a^2 - c_1 b^2}{a\,(a^2 + b^2)}, \frac{b(c_1 - c_2)}{a^2 + b^2}\right). \tag{6.32}$$

Finally, we use (6.26) to obtain from (6.30) and (6.32) that the distance between the lines l_1 and l_2 is

$$|P_1 Q_1| = \frac{|c_1 - c_2|}{(a^2 + b^2)^{1/2}}. \tag{6.33}$$

It is easily verified that (6.33) also holds when $a = 0$.

We can derive (6.33) in a more interesting way. First, let us assume that l_1 and l_2 are not parallel to the x-axis or the y-axis. In Figure 6.20 the parallel lines l_1 and l_2 cut the x-axis at the two points P_1 and P_2, respectively. Let $|P_1 P_2| = X$, and let us write $d = |P_1 Q_1|$, the distance between the lines l_1 and l_2. It then follows from triangle $Q_1 P_1 P_2$ that

$$\frac{|P_1 Q_1|}{|P_1 P_2|} = \frac{d}{X} = \sin \theta, \tag{6.34}$$

where θ is the angle that each line l_1 and l_2 makes with the x-axis. We can see that each line l_1 and l_2 makes an angle $\frac{\pi}{2} - \theta$ with the y-axis. Then if Y denotes the distance between the two points where the lines l_1 and l_2 cut the y-axis, we similarly obtain

$$\frac{d}{Y} = \sin \left(\frac{\pi}{2} - \theta\right) = \cos \theta. \tag{6.35}$$

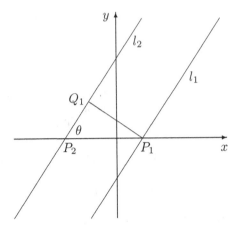

FIGURE 6.20. $P_1Q_1 = d$ is the distance between the parallel lines l_1 and l_2.

Since $\cos^2 \theta + \sin^2 \theta = 1$, we deduce from (6.34) and (6.35) that

$$\frac{d^2}{X^2} + \frac{d^2}{Y^2} = 1,$$

which we can write in the form

$$\frac{1}{d^2} = \frac{1}{X^2} + \frac{1}{Y^2},$$

so that

$$d = \frac{XY}{(X^2 + Y^2)^{1/2}}. \tag{6.36}$$

Now let l_1 and l_2 be represented by the equations in (6.29). If $a \neq 0$, the lines l_1 and l_2 cut the x-axis where $x = -c_1/a$ and $x = -c_2/a$, respectively. If $b \neq 0$, l_1 and l_2 cut the y-axis where $y = -c_1/b$ and $y = -c_2/b$. Thus when $a \neq 0$ and $b \neq 0$ we obtain

$$X = \left| \frac{c_1 - c_2}{a} \right| \quad \text{and} \quad Y = \left| \frac{c_1 - c_2}{b} \right|,$$

and if we substitute these values into (6.36), we obtain (6.33). ∎

Let the point A_j have coordinates (x_j, y_j), for $j = 1, 2,$ and 3. Then the point C_1 with coordinates $(\frac{1}{2}(x_2 + x_3), \frac{1}{2}(y_2 + y_3))$ is the midpoint of $A_2 A_3$. (See Example 6.4.2.) If we take λ times the coordinates of A_1 plus $1 - \lambda$ times the coordinates of C_1, we get a point on the median $A_1 C_1$. In particular, the choice of $\lambda = \frac{1}{3}$ gives the point

$$C = \left(\tfrac{1}{3}(x_1 + x_2 + x_3), \tfrac{1}{3}(y_1 + y_2 + y_3) \right). \tag{6.37}$$

We can show in the same way that C lies also on the other two medians, A_2C_2 and A_3C_3. Thus coordinate geometry has given us a simple proof that the medians of a triangle are concurrent at the centroid C, and that C trisects each of the medians.

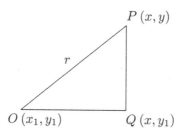

FIGURE 6.21. Derivation of the equation of a circle.

To obtain the equation of a circle, consider the triangle OPQ in Figure 6.21, where O is the fixed point with coordinates (x_1, y_1), P is any point (x, y) whose distance from O is r, and Q is the point where the vertical line through P intersects the horizontal line through O. Although Figure 6.21 shows P as lying above and to the right of O, it can lie anywhere on the circle with center at O and with radius r. Then, since the triangle OPQ has a right angle at Q, it follows from Pythagoras's theorem that $|OQ|^2 + |PQ|^2 = |OP|^2$. Since $|OQ|^2 = (x - x_1)^2$, $|PQ|^2 = (y - y_1)^2$, and $|OP|^2 = r^2$, we obtain

$$(x - x_1)^2 + (y - y_1)^2 = r^2. \tag{6.38}$$

The circle with center at (x_1, y_1) and radius r is the set of all points (x, y) whose coordinates satisfy (6.38). Thus the coordinate geometry of Descartes allows us to turn geometrical problems into algebraic problems.

Let us find the coordinates of the circumcenter A of the triangle $A_1A_2A_3$, where A_j has coordinates (x_j, y_j), for $j = 1, 2$, and 3. We can assume that y_1, y_2, and y_3 are all different. For if two of these three numbers were equal, we could rotate the triangle so that the new y coordinates were all different. Since the line A_2A_3 has gradient $(y_2 - y_3)/(x_2 - x_3)$, any line perpendicular to A_2A_3 has gradient $-(x_2 - x_3)/(y_2 - y_3)$. The equation of the line that passes through C_1, the midpoint of A_2A_3, and is perpendicular to A_2A_3 is

$$\frac{y - \frac{1}{2}(y_2 + y_3)}{x - \frac{1}{2}(x_2 + x_3)} = -\left(\frac{x_2 - x_3}{y_2 - y_3}\right). \tag{6.39}$$

Similarly, the equation of the line that passes through C_2, the midpoint of A_3A_1, and is perpendicular to A_3A_1 is

$$\frac{y - \frac{1}{2}(y_3 + y_1)}{x - \frac{1}{2}(x_3 + x_1)} = -\left(\frac{x_3 - x_1}{y_3 - y_1}\right). \tag{6.40}$$

Notice that we can change (6.39) into (6.40) by changing the suffixes 1, 2, and 3 in *cyclic* order, so that 1 becomes 2, 2 becomes 3, and 3 becomes 1. If we express the line in (6.39) in the form

$$y = m_1 x + d_1, \tag{6.41}$$

we find that

$$m_1 = -\left(\frac{x_2 - x_3}{y_2 - y_3}\right), \quad d_1 = \frac{1}{2}(y_2 + y_3) + \frac{(x_2 + x_3)(x_2 - x_3)}{2(y_2 - y_3)}. \tag{6.42}$$

By permuting the suffixes in (6.41) and (6.42) cyclically, we can recast the equation of the straight line in (6.40) in the form

$$y = m_2 x + d_2, \tag{6.43}$$

where

$$m_2 = -\left(\frac{x_3 - x_1}{y_3 - y_1}\right), \quad d_2 = \frac{1}{2}(y_3 + y_1) + \frac{(x_3 + x_1)(x_3 - x_1)}{2(y_3 - y_1)}. \tag{6.44}$$

Let (x_A, y_A) denote the coordinates of the point where the lines defined by (6.41) and (6.43) intersect. We see that

$$m_1 x_A + d_1 = m_2 x_A + d_2,$$

so that

$$x_A = -\left(\frac{d_1 - d_2}{m_1 - m_2}\right). \tag{6.45}$$

We now wish to express x_A directly in terms of the coordinates of the vertices of the triangle $A_1 A_2 A_3$. For the denominator of the fraction on the right of (6.45), we obtain from (6.42) and (6.44) that

$$m_1 - m_2 = \frac{-\Delta_{xy}}{(y_2 - y_3)(y_3 - y_1)}, \tag{6.46}$$

where

$$\Delta_{xy} = x_1(y_2 - y_3) + x_2(y_3 - y_1) + x_3(y_1 - y_2). \tag{6.47}$$

Note that Δ_{xy} must be nonzero, since the gradients m_1 and m_2 are different. For the numerator on the right of (6.45), a little work shows that

$$d_2 - d_1 = \frac{\Delta_y - \{x_1^2(y_2 - y_3) + x_2^2(y_3 - y_1) + x_3^2(y_1 - y_2)\}}{2(y_2 - y_3)(y_3 - y_1)}, \tag{6.48}$$

where

$$\Delta_y = (y_2 - y_3)(y_3 - y_1)(y_1 - y_2). \tag{6.49}$$

Then, on combining our results in (6.45), (6.46), and (6.48), we find that the x coordinate x_A of the circumcenter A of the triangle $A_1A_2A_3$ is

$$x_A = \frac{\{x_1^2(y_2 - y_3) + x_2^2(y_3 - y_1) + x_3^2(y_1 - y_2)\} - \Delta_y}{2\Delta_{xy}}. \qquad (6.50)$$

The y coordinate, y_A, obtained by interchanging x and y in (6.50), is

$$y_A = \frac{\{y_1^2(x_2 - x_3) + y_2^2(x_3 - x_1) + y_3^2(x_1 - x_2)\} - \Delta_x}{2\Delta_{yx}}. \qquad (6.51)$$

Notice that the coordinates x_A and y_A, which we derived by finding the point of intersection of the perpendicular bisectors of A_2A_3 and A_3A_1, are symmetric in x_1, x_2, and x_3, and also in y_1, y_2, and y_3. Thus we would obtain the same point (x_A, y_A) as the point of intersection of the perpendicular bisectors of A_3A_1 and A_1A_2. This confirms algebraically what we found (much more easily) in Construction 10, that the perpendicular bisectors of the sides of a triangle are concurrent.

Using the same approach, we can similarly find the point of intersection of the perpendiculars from A_1 to A_2A_3 and A_2 to A_3A_1. Let this point have coordinates (x_B, y_B). We find that

$$x_B = \frac{\Delta_y - \{x_2x_3(y_2 - y_3) + x_3x_1(y_3 - y_1) + x_1x_2(y_1 - y_2)\}}{\Delta_{xy}}, \qquad (6.52)$$

and y_B, obtained by interchanging x and y in (6.52), is

$$y_B = \frac{\Delta_x - \{y_2y_3(x_2 - x_3) + y_3y_1(x_3 - x_1) + y_1y_2(x_1 - x_2)\}}{\Delta_{yx}}. \qquad (6.53)$$

We observe that the coordinates (x_B, y_B), like (x_A, y_A), are symmetric in x_1, x_2, and x_3, and also in y_1, y_2, and y_3. This *proves* that the three perpendiculars from the vertices of a triangle to the opposite sides are concurrent, and the equations (6.52) and (6.53) give the coordinates of the point of concurrence B, the orthocenter of the triangle $A_1A_2A_3$.

A comparison of (6.50) and (6.52) prompts the observation that we can eliminate Δ_y by adding twice x_A to x_B. On carrying out this calculation, we obtain

$$2x_A + x_B = \frac{X_1(y_2 - y_3) + X_2(y_3 - y_1) + X_3(y_1 - y_2)}{\Delta_{xy}}, \qquad (6.54)$$

where

$$X_1 = x_1^2 - x_2x_3, \quad X_2 = x_2^2 - x_3x_1, \quad X_3 = x_3^2 - x_1x_2.$$

We can express

$$X_j = x_j(x_1 + x_2 + x_3) - (x_2x_3 + x_3x_1 + x_1x_2), \quad \text{for } 1 \le j \le 3,$$

and we note that

$$(x_2x_3 + x_3x_1 + x_1x_2)\{(y_2 - y_3) + (y_3 - y_1) + (y_1 - y_2)\} = 0,$$

since the second factor on the left of the latter equation is zero. Thus the numerator on the right of (6.54) can be recast in the form

$$(x_1 + x_2 + x_3)\{x_1(y_2 - y_3) + x_2(y_3 - y_1) + x_3(y_1 - y_2)\}.$$

Since

$$x_1(y_2 - y_3) + x_2(y_3 - y_1) + x_3(y_1 - y_2) = \Delta_{xy},$$

(6.54) greatly simplifies to give

$$2x_A + x_B = x_1 + x_2 + x_3.$$

We can treat the y coordinates similarly, and so obtain

$$\frac{2}{3}x_A + \frac{1}{3}x_B = x_C, \quad \frac{2}{3}y_A + \frac{1}{3}y_B = y_C, \tag{6.55}$$

say, where $x_C = \frac{1}{3}(x_1+x_2+x_3)$ and $y_C = \frac{1}{3}(y_1+y_2+y_3)$ are the coordinates of the centroid of the triangle. We have shown that the circumcenter A, the orthocenter B, and the centroid C all lie on the same straight line, called the Euler line, and that C divides the line segment AB in the ratio $1 : 2$; that is, C is one-third of the way from A to B. Although these properties of the points A, B, and C were not *proved* until the eighteenth century, it is difficult to believe that they were not known empirically to at least some individuals among the many practitioners of the art of ruler and compass constructions two thousand years earlier.

Our verification of the above facts concerning the Euler line illustrates a general point about coordinate geometry. Although it can be aesthetically pleasing to use, it sometimes leaves us none the wiser about the underlying mathematics. It can seem as mechanical in its operation as a ruler and compass construction. Yet, and this shows its importance, it can provide us with *proofs* of geometrical results.

Fermat showed (see Edwards [7]) that if an equation of the form

$$ax^2 + bxy + cy^2 + dx + ey + f = 0, \tag{6.56}$$

is satisfied by any point with Cartesian coordinates (x, y), then it is the equation of an ellipse, hyperbola, parabola, or pair of straight lines. Thus, just as these curves are unified geometrically in that they are all sections of a cone, they are also unified algebraically by (6.56).

Problem 6.4.1 Consider a triangle ABC. Let D denote the midpoint of BC, and let D, A, B, and C have coordinates $(0,0)$, (x_1, y_1), $(-x_2, 0)$, and $(x_2, 0)$, respectively. Hence show that $|AB|^2 + |AC|^2 = 2(|AD|^2 + |BD|^2)$, which we obtained using other methods in Problem 1.2.5.

Problem 6.4.2 Verify that $\Delta_{yx} = -\Delta_{xy}$, where Δ_{xy} is defined in (6.47), and Δ_{yx} is obtained by interchanging x and y in (6.47).

Problem 6.4.3 Verify that the equation of the straight line that is parallel to the line l with equation $ax + by + c = 0$ and passes through the point $P = (x_1, y_1)$ is $ax + by - ax_1 - by_1 = 0$. Deduce from (6.33) that the distance d from the point P to the line l is given by

$$d = \frac{|ax_1 + by_1 + c|}{(a^2 + b^2)^{1/2}}.$$

Problem 6.4.4 Consider the equation

$$\frac{x^2}{a^2} + \frac{y^2}{b^2} = 1,$$

with $a > b > 0$. Write $x = au/b$, and $y = v$, and verify that the above equation becomes

$$u^2 + v^2 = b^2,$$

the equation of a circle with center $u = 0$, $v = 0$, and radius b. Thus the first equation defines a curve in the coordinates x and y that is a circle "stretched" in the x direction. This is an ellipse.

6.5   Regular Polyhedra

In Section 6.1 we discussed regular polygons. The analogue of a polygon in three dimensions is a *polyhedron*. Just as a polygon is constructed by fixing line segments together, a polyhedron is constructed by fixing polygons together. A polyhedron that is constructed by fixing together a number of copies of the same *regular* polygon in a fully symmetric way is called a *regular polyhedron*. The plural of polyhedron is *polyhedra*, and the regular polyhedra are also called the Platonic solids. These are discussed in Euclid's *Elements*, and the best known regular polyhedron is the *cube*, which has six square *faces*, twelve *edges*, and eight *vertices*. Figure 6.22 displays a cube on the left and its *net*, the cross-shaped diagram in the middle of Figure 6.22, constructed from six squares. We can cut out this shape from a piece of cardboard and fold it to make a cube. (It is useful to augment the net by adding some extra pieces to help hold the cube together. We can coat the extra pieces with glue and fold them, out of sight, behind faces of the cube.) Coxeter [4] states that Leonardo da Vinci (1452–1519) made skeletal models of polyhedra, using strips of wood for their edges and leaving their faces to be imagined. When a model of this kind is viewed from just outside the center of one face, this face is seen as a large polygon

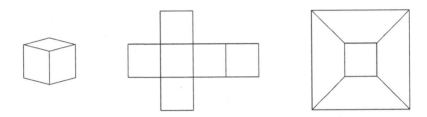

FIGURE 6.22. The cube, its net, and its Schlegel diagram.

with all the other faces filling its interior. Such a view of the cube is shown on the right of Figure 6.22. It is called a *Schlegel* diagram.

We will use the notation $\{n, m\}$ to denote a regular polyhedron that is constructed from polygons with n sides, with m such polygons meeting around each vertex. Note that we must have $n \geq 3$ and $m \geq 3$. In this notation the cube is denoted by $\{4, 3\}$. As we saw in (3.34), each angle of a regular polygon with n sides is $(1 - 2/n)\pi$. This angle increases with n and has the value $2\pi/3$ when $n = 6$. Thus three regular hexagons fit together precisely, lying flat on the plane, as shown in Figure 6.23.

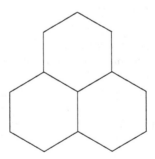

FIGURE 6.23. The plane can be covered with hexagons.

It follows that there is no regular polyhedron constructed from polygons with six or more sides. Incidentally, Figure 6.23 shows that the whole plane can be covered by regular hexagons of the same size. We can express such a covering of the plane by the notation $\{6, 3\}$. It is easy to verify that there are only two other regular polygons that can be used to cover the plane, namely the equilateral triangle and the square, and we will denote these coverings by $\{3,6\}$ and $\{4,4\}$, respectively. Coverings of the plane, such as $\{6,3\}$, $\{3,6\}$, and $\{4,4\}$, are also called plane *tessellations* or *tilings*.

The above analysis shows that the only possible regular polyhedra are those that are constructed from equilateral triangles, squares, and regular pentagons. First, let us consider regular polyhedra that are constructed from equilateral triangles, whose angles are all $\pi/3$. The only possible regular polyhedra $\{3, m\}$ are those for which

$$m \cdot \frac{\pi}{3} < 2\pi, \quad \text{with} \quad m \geq 3,$$

which gives $m = 3, 4$ or 5. Second, when $n = 4$, we need to consider regular polyhedra that are constructed from squares, whose angles are all $\pi/2$. The only possible regular polyhedra $\{4, m\}$ are those for which

$$m \cdot \frac{\pi}{2} < 2\pi, \quad \text{with} \quad m \geq 3.$$

The only solution is $m = 3$, giving the cube, which we have already studied in Figure 6.22. The only other possible regular polyhedra are those of the form $\{5, m\}$, involving regular pentagons, whose angles (see (3.34)) are all $3\pi/5$. In this case we require that

$$m \cdot \frac{3\pi}{5} < 2\pi, \quad \text{with} \quad m \geq 3,$$

and the only solution is $m = 3$. Thus there are only *five* possible regular polyhedra, namely

$$\{3,3\}, \quad \{3,4\}, \quad \{3,5\}, \quad \{4,3\} \quad \{5,3\}. \tag{6.57}$$

We know that the $\{4,3\}$ case does indeed give a regular polyhedron (the cube), and we will now follow up the other four possibilities defined in (6.57). As we will see, all five of the above configurations $\{n, m\}$ yield regular polyhedra.

FIGURE 6.24. The regular tetrahedron, its net, and its Schlegel diagram.

Let us call a regular polygon with n vertices an n-gon. Consider any of the cases $\{n, m\}$ defined in (6.57). We begin by fixing together m n-gons so that they meet at a vertex. This creates further vertices. If possible, we

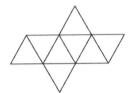 

FIGURE 6.25. The regular octahedron, its net, and its Schlegel diagram.

add more n-gons, so that there are m at each new vertex. We continue this process to see whether, ultimately, we do obtain a regular polyhedron $\{n, m\}$. The reader may find it helpful to apply this constructive method to obtain the cube, since it is already familiar to us.

For the case $\{3, 3\}$, we begin by fixing together three equilateral triangles around a point P. There are now three other vertices, say A, B, and C. The three edges BC, CA, and AB are all equal, and we can fix a fourth equilateral triangle to ABC to complete a regular polyhedron. This is called the *regular tetrahedron* which, together with its net and its Schlegel diagram, is depicted in Figure 6.24. The regular tetrahedron can also be described, with less precision, as a pyramid on a triangular base.

Let us pursue the second case in (6.57), namely $\{3, 4\}$. We begin by joining together four equilateral triangles around a point P. There are now four other vertices, say A, B, C, and D. By symmetry, the vertices A, B, C, and D must lie in the same plane in a square. Thus the configuration $PABCD$ is a pyramid on a square base. Following the procedure described above, we can complete the polyhedron that is displayed in Figure 6.25, together with its net and Schlegel diagram. This is the *octahedron*, whose name means that it has eight faces. An octahedron can also be constructed by gluing together two copies of the pyramid $PABCD$.

Let us consider the cube again, and join the center of each face to the centers of all four neighboring faces. This construction gives an octahedron whose vertices are the centers of the six faces of the cube. Conversely, if we join the center of each face of an octahedron to the centers of all three neighboring faces, we obtain a cube. The vertices of the cube are the centers of the faces of the octagon. Thus there is a correspondence between the faces of a cube and the vertices of an octahedron, and a correspondence between the vertices of a cube and the faces of an octahedron. We say that each of these two polyhedra is the *dual* of the other. On applying the same process to the tetrahedron, joining the center of each face to the centers of all neighboring faces (that is, to *all* faces), we obtain another tetrahedron. We say that the tetrahedron is its own dual.

  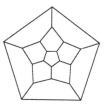

FIGURE 6.26. The regular dodecahedron, its net, and its Schlegel diagram.

The two remaining configurations defined in (6.57) are the most glorious. The polyhedron $\{5, 3\}$, involving the regular pentagon, is depicted in Figure 6.26, together with its net and its Schlegel diagram. It is called the regular dodecahedron, meaning twelve faces. The polyhedron $\{3, 5\}$, with five equilateral triangles around each vertex, is shown in Figure 6.27, together with its net and its Schlegel diagram. This is the regular icosahedron, meaning twenty faces. Note that the dodecahedron and icosahedron are duals. The net in Figure 6.27 illustrates the fact that $\{3, 6\}$ provides a tessellation of the plane, to which we referred above. Indeed, the whole net itself in Figure 6.27 provides a tessellation of the plane.

The basic data about the regular polyhedra are summarized in Table 6.1, where F, E, and V denote the number of faces, edges, and vertices, respectively. Note that the value of F for a given regular polyhedron is the same as the value of V for its dual, and the dual of $\{n, m\}$ is $\{m, n\}$. Using the appropriate ruler and compass constructions in Section 6.1, we can construct the nets for the regular polyhedra. The reader will find it instructive and satisfying to make models of all five regular polyhedra.

Name	$\{n, m\}$	F	E	V
tetrahedron	$\{3, 3\}$	4	6	4
cube	$\{4, 3\}$	6	12	8
octahedron	$\{3, 4\}$	8	12	6
dodecahedron	$\{5, 3\}$	12	30	20
icosahedron	$\{3, 5\}$	20	30	12

TABLE 6.1. The five Platonic solids.

From the definition of duality given above, the dual of a given Platonic solid is found on interchanging the values of F and V in Table 6.1. Thus, as already stated, the tetrahedron is its own dual, the cube and the octahedron are duals, and the dodecahedron and the icosahedron are duals.

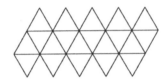 

FIGURE 6.27. The regular icosahedron, its net, and its Schlegel diagram.

In Section 3.3 we considered the star of Pythagoras, or pentagram, which can be constructed by extending the sides of a regular pentagon. We can do this for any polygon with five or more sides to obtain star-like figures. For example, if we begin with a regular hexagon, we obtain the star of David. Analogously, we can obtain star-like polyhedra if we extend faces of polyhedra. These are called *stellated* polyhedra, *stella* being the Latin word for star. It is obvious that we cannot stellate the tetrahedron or the cube. However, we can construct a *stellated octahedron*. Although this figure was known much earlier, it was studied by Johannes Kepler (1571–1630). It consists of a regular octahedron with a tetrahedron glued to each face, and can be viewed as two intersecting tetrahedra. Of the many known stellated polyhedra, I will mention only two more, both of which were also studied by Kepler, and are derived from the regular dodecahedron and regular icosahedron. To obtain the first of these, we begin with a regular dodecahedron and extend the edges around each face to give a pentagram. This gives the figure called the *small stellated dodecahedron*. To obtain the other figure we begin with the regular icosahedron, and consider any vertex P. We then take the pentagon formed from the five neighboring vertices of P and replace it by a pentagram. If we do this for each of the 12 vertices P, we obtain the figure that is called the *great stellated dodecahedron*.

In Book V of his *Mathematical Collection*, Pappus of Alexandria discusses a set of polyhedra that, like the regular polyhedra, are constructed from regular polygons. In this case, the polygons do not all have the same number of sides. However, there is the same configuration of polygons around each vertex. These are called the *semiregular* polyhedra. Pappus attributes their discovery to Archimedes, and they are also called the Archimedean solids. We denote a semiregular polyhedron by $(n_1, n_2, \ldots, n_m)$, where n_1, $n_2, \ldots, n_m$, not all equal, denote the number of sides in the polygons, taken in cyclic order around each vertex. For example, $(3, 4, 3, 4)$ denotes the semiregular polyhedron that has four regular polygons, with sides 3, 4, 3, and 4, around each vertex. This is called the *cuboctahedron*. It is obtained by chopping off each of the eight corners of a cube by joining together the midpoints of adjoining edges. If we chop off the corners of the cube so that each of its six faces becomes a regular octagon, we obtain the polyhedron $(3, 8, 8)$, called the *truncated cube*. We could also use this notation to describe the regular polyhedra and tessellations of the plane. For example, $(4, 4, 4)$ denotes the cube, while $(3, 3, 3, 3, 3, 3)$, $(4, 4, 4, 4)$, and $(6, 6, 6)$ denote the tessellations of the plane described above.

For any Archimedean solid $(n_1, n_2, \ldots, n_m)$, the condition that the sum of the angles around each vertex must be less than 2π yields the inequality

$$\sum_{j=1}^{m} \left(1 - \frac{2}{n_j} \right) \pi < 2\pi, \quad \text{where} \quad m \geq 3 \quad \text{and every} \quad n_j \geq 3,$$

and we can deduce that this is equivalent to

$$\sum_{j=1}^{m} \frac{1}{n_j} > \frac{1}{2}m - 1, \quad \text{where} \quad m \geq 3 \quad \text{and every} \quad n_j \geq 3. \tag{6.58}$$

The inequality (6.58) also applies to the regular polyhedra, where each $n_j = n$. In this case we find that (6.58) is equivalent to the identity

$$\frac{1}{m} + \frac{1}{n} > \frac{1}{2}, \quad \text{where} \quad m, n \geq 3. \tag{6.59}$$

In our discussion of the regular polyhedra, we effectively worked through the inequality (6.59) for the cases $n = 3$, $n = 4$, and $n = 5$ separately, and found that every feasible solution of (6.59) yields a regular polyhedron. In contrast, not all solutions of (6.58) correspond to semiregular polyhedra. For example, with $m = 3$, $n_1 = n_2 = 3$ and $n_3 = n$ in (6.58), we find that although the inequality holds for all values of n, none of the solutions with $n > 3$ yields a polyhedron. I state without proof that of all the solutions of (6.58) where the n_j are not all equal, only *thirteen* solutions yield polyhedra. All thirteen Archimedean solids $(n_1, n_2, \ldots, n_m)$ are listed in Table 6.2, together with their names and numbers of faces, edges, and vertices. The number of faces of each of the different constituent polygons is given in square brackets after the number of faces in column F. For example, the truncated cube has 14 faces (8 triangles and 6 octagons). The Archimedean solid with the smallest number of faces (4 triangles and 4 hexagons) is the truncated tetrahedron. This is obtained by taking a regular tetrahedron and removing four small regular tetrahedra, each including a vertex of the large tetrahedron and whose edges are one-third of the length of the edges of the large tetrahedron. The design of many soccer balls in current use is based on the truncated icosahedron, $(5, 6, 6)$, which has 12 pentagons and 20 hexagons. The Archimedean solid with most faces is the snub dodecahedron, with 80 triangular and 12 pentagonal faces. The great rhombicosidodecahedron is the Archimedean solid with the most edges and the most vertices, and it also has the longest name! Making models of all the Archimedean solids would make a worthy class project.

$(n_1, n_2, \ldots, n_m)$, name	F	E	V
$(3, 6, 6)$, truncated tetrahedron	$8\,[4, 4]$	18	12
$(3, 8, 8)$, truncated cube	$14\,[8, 6]$	36	24
$(3, 10, 10)$, truncated dodecahedron	$32\,[20, 12]$	90	60
$(4, 6, 6)$, truncated octahedron	$14\,[6, 8]$	36	24
$(4, 6, 8)$, great rhombicuboctahedron	$26\,[12, 8, 6]$	48	24
$(4, 6, 10)$, great rhombicosidodecahedron	$62\,[30, 20, 12]$	180	120
$(5, 6, 6)$, truncated icosahedron	$32\,[12, 20]$	90	60
$(3, 4, 3, 4)$, cuboctahedron	$14\,[8, 6]$	24	12
$(3, 5, 3, 5)$, icosidodecahedron	$32\,[20, 12]$	60	30
$(3, 4, 4, 4)$, small rhombicuboctahedron	$26\,[8, 18]$	48	24
$(3, 4, 5, 4)$, small rhombicosidodecahedron	$62\,[20, 30, 12]$	120	60
$(3, 3, 3, 3, 4)$, snub cube	$38\,[32, 6]$	60	24
$(3, 3, 3, 3, 5)$, snub dodecahedron	$92\,[80, 12]$	150	60

TABLE 6.2. The thirteen Archimedean solids.

We can also look for solutions of (6.58) where the inequality is replaced by an equality. Some of these solutions, but not *all*, give tessellations of the plane. I state without proof that there are eight of these. They are called Archimedean tessellations and are listed in Table 6.3. The last two both have 3 triangles and 2 squares around each vertex, but taken in a different order. Of all eight tessellations, my favorite is $(3, 3, 4, 3, 4)$.

$$(3, 12, 12) \qquad (4, 6, 12) \qquad (4, 8, 8) \qquad (3, 6, 3, 6)$$
$$(3, 4, 6, 4) \qquad (3, 3, 3, 3, 6) \qquad (3, 3, 3, 4, 4) \qquad (3, 3, 4, 3, 4)$$

TABLE 6.3. The eight Archimedean tessellations.

There is an infinite number of ways of tessellating the plane, and we have looked only at some of those that involve regular polygons. Let us call any shape that can be used repeatedly to tessellate the plane a *motif*. We can obviously use any parallelogram as a motif, or a cross constructed from five squares, or a triangle of any shape. The last one is obvious, because we can put two identical triangles together to make a parallelogram. However, it is not so obvious that any quadrilateral can be used as a motif. We can find motifs for the Archimedean tessellations. Motifs for $(3, 3, 4, 3, 4)$ and $(3, 4, 6, 4)$ are given in Figure 6.28.

Let us begin with the tessellation $(4, 4, 4, 4)$. It is obvious that we can amend every square in the same way so that we still have a tessellation of the plane. A very simple example is given in Figure 6.29.

A segment has been removed from the left side of the square and added to the right side. If I had chosen an appropriate triangle as the amending segment in Figure 6.29, I could have changed the square into a parallelogram.

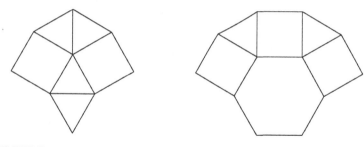

FIGURE 6.28. Motifs for the tessellations $(3,3,4,3,4)$ and $(3,4,6,4)$.

We can also amend the top and bottom edges of the square. Similarly, we can create amended versions of other well-known tessellations, such as $(3,3,3,3,3,3)$. This process is one of the keys to understanding the many very beautiful and fascinating tessellations of the plane that were created by the artist and mathematician M. C. Escher (1898–1972).

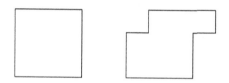

FIGURE 6.29. An amended square that still tessellates the plane.

Roger Penrose (born 1931) invented tessellations that are constructed from two particular quadrilaterals, called a kite and a dart, and involve the golden section. The kite and dart are obtained by dissecting the rhombus $ABCD$ in Figure 6.30. The sides of the rhombus are all of length $\frac{1}{2}\left(\sqrt{5}+1\right)$, $|AE|=|EC|=|BE|=1$, and the angle ABC is $2\pi/5$.

It can be proved that the kite and dart can be used to cover the plane in an infinite number of ways that are not periodic. This means that if we were able to make a transparency of such a tessellation, there is no way we could move it, without rotating it, so that it matched the tessellation. More material on tessellations can be found in Wells [17].

If we examine Table 6.2 we observe that for every Archimedean solid, F plus V is approximately equal to E. More precisely, we see that

$$F - E + V = 2, \tag{6.60}$$

and an inspection of Table 6.1 reveals that (6.60) holds also for the five Platonic solids. Although the identity (6.60) is often named after Euler and Descartes, it is hard to believe that it was not known to Pappus of

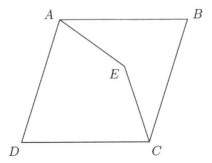

FIGURE 6.30. The rhombus $ABCD$ dissected into Penrose's kite and dart.

Alexandria and to Archimedes. If we begin with a sphere, choose $n \geq 1$ vertices on its surface, and join them up in any way that creates a map of polygonal-shaped countries, then the number of faces, edges, and vertices always satisfies (6.60), and will continue to do so if we stretch or contract any of the line segments. In this way we obtain any polyhedron that is said to be *simply connected*, meaning that it has no holes in it. Beginning with such a polyhedron, let us stretch it, pull the assemblage of faces, edges, and vertices off the sphere and flatten it out to give a map of polygons on the plane that has same number of vertices and edges as before, but has one face fewer. The resulting map will be equivalent to the Schlegel diagram in the case of the Platonic solids.

Let us now consider a map of this kind. We will show that its number of faces, edges, and vertices satisfies the equation $F - E + V = 1$. Since we "lost" a face when we removed the assemblage from the sphere, this is equivalent to proving that (6.60) holds for any polyhedron. To construct such a diagram, we will begin with the empty plane and place one vertex on it. At this stage we have no faces, no edges, and one vertex, and thus $F - E + V = 1$. We then build up the required map by adding one edge at a time. Unless we are completing a polygon, each time we add one edge, we add one vertex. However, if we complete a polygon by adding an edge, we add one face. In either case, we do not change the value of $F - E + V$ with the addition of an edge, and this proves that $F - E + V = 1$ when we have completed the construction of the chosen map. We have thus proved the following theorem.

Theorem 6.5.1 If F, E, and V denote the number of faces, edges, and vertices of a given simply connected polyhedron, then $F - E + V = 2$. ∎

For each Platonic solid $\{n, m\}$ we see that if we cut out every face, we would have nF edges. However, each edge of the Platonic solid connects two faces, and so $nF = 2E$. Similarly, each vertex is on m edges, and each edge connects two vertices. It follows that $mV = 2E$, and thus we have

$$nF = 2E = mV. \tag{6.61}$$

Beginning with (6.61) and the identity $F - E + V = 2$ we can derive explicit values of F, E, and V for the platonic solids. See Problem 6.5.6.

Problem 6.5.1 Show that an octahedron and four tetrahedra can be fixed together to form a larger tetrahedron, and that three-dimensional space can be filled using tetrahedra and octahedra.

Problem 6.5.2 Cut out copies of the appropriate regular polygons and explore the eight Archimedean tessellations given in Table 6.3.

Problem 6.5.3 Construct a dual of the $(3, 3, 4, 3, 4)$ tessellation by joining the centers of adjacent polygons. This is called the Cairo tessellation. Observe that it has a pentagonal motif that has four sides of one length and one shorter side.

Problem 6.5.4 Begin with the $(3, 3, 4, 3, 4)$ tessellation whose triangles and squares have sides of length 1, and construct its dual, as in Problem 6.5.3. Show that the pentagon that occurs in the Cairo tessellation has sides, taken in order, whose lengths are a, a, b, a, and b, where $a = \frac{1}{2}$ and $b = \frac{1}{\sqrt{2}}$. Show also that the angle enclosed by the two adjacent sides of length $\frac{1}{2}$ is $\frac{\pi}{3}$, and that the four other angles are $\frac{5\pi}{12}$.

Problem 6.5.5 Show that the dual tessellation of every Archimedean tessellation is composed of repetitions of the same polygon.

Problem 6.5.6 Use (6.61) to express E and V in terms of F and so deduce from (6.60) that

$$F = \frac{4m}{2m + 2n - mn}.$$

Show also that

$$E = \frac{2mn}{2m + 2n - mn},$$

and

$$V = \frac{4n}{2m + 2n - mn}.$$

Note that interchanging m and n in the above expressions corresponds to interchanging F and V, and leaves E unchanged, as we should expect from duality.

Problem 6.5.7 Deduce from the expressions obtained for F, E, and V in Problem 6.5.6 that $2m + 2n - mn > 0$. Show that the latter inequality is equivalent to (6.59), and is also equivalent to

$$(m - 2)(n - 2) < 4.$$

Show that since we must have $m \geq 3$ and $n \geq 3$, this last inequality has only five solutions and that these correspond to the five Platonic solids.

Problem 6.5.8 For the Platonic solid $\{n, m\}$, let δ denote the amount by which the sum of the angles at a vertex falls short of 2π. (Thus $\delta = \pi$ for the tetrahedron and $\delta = \pi/5$ for the dodecahedron.) Show that

$$\delta = 2\pi - m \left(1 - \frac{2}{n} \right) \pi,$$

and deduce that $\delta = 4\pi/V$, where V is the number of vertices in the Platonic solid.

7
The Algebra of Groups

The mathematician's patterns, like the painter's or the poet's, must be beautiful; the ideas, like the colours or the words, must fit together in a harmonious way.

G. H. Hardy (1877–1947)

7.1 Introduction

My dictionary defines *algebra* as "the branch of mathematics that uses letters to represent numbers and other quantities in formulas and equations." In terms of this broad definition we have used algebra in every chapter of this book, beginning on page 3 when we denoted the nth triangular number by T_n, and continuing with the application of algebraic methods to discuss integers, rational numbers, real numbers, and complex numbers. If we have two quantities a and b in any of these number systems, we take it for granted that a and b *commute*, meaning that

$$ab = ba, \tag{7.1}$$

and that other algebraic relations are satisfied, such as

$$a(b + c) = ab + ac. \tag{7.2}$$

The relation $ab = ba$ is called the *commutative* property, and the second equation, $a(b+c) = ab+ac$, describes the *distributive* property. In the nineteenth century, new algebraic systems emerged that were not commutative, meaning that (7.1) does not always hold. Among those whose pioneering work transformed algebra in this way are William Rowan Hamilton (1805–1865), H. G. Grassmann (1809–1877), and Arthur Cayley (1821–1895).

Cayley's contribution to these exciting advances in algebra involves the *matrices* that occur in linear transformations. Given a system of Cartesian coordinates in x and y (see Section 6.4), let us create a new Cartesian system involving coordinates x' and y', and let the new coordinates be connected to the old coordinates by the equations

$$x' = ax + by,$$
$$y' = cx + dy,$$

where $a, b, c,$ and d are real numbers. For example, if $a = d = 0$ and $-b = c = 1$, the point (x, y) is transformed to $(x', y') = (-y, x)$, and you should be able to see by drawing a diagram that the above transformation is equivalent to rotating the (x, y) plane counterclockwise through a right angle. To emphasize its dependence on a, b, c, and d, the above transformation is written in the form

$$\begin{bmatrix} x' \\ y' \end{bmatrix} = \begin{bmatrix} a & b \\ c & d \end{bmatrix} \begin{bmatrix} x \\ y \end{bmatrix}, \tag{7.3}$$

and the entity involving a, b, c, and d, contained within left and right brackets in (7.3), is called a 2×2 *matrix*. Let us now carry out a second transformation in which the system involving x' and y' is transformed into one involving x'' and y'', such that

$$\begin{bmatrix} x'' \\ y'' \end{bmatrix} = \begin{bmatrix} a' & b' \\ c' & d' \end{bmatrix} \begin{bmatrix} x' \\ y' \end{bmatrix}. \tag{7.4}$$

It is now a simple matter to obtain the direct connection between the (x'', y'') system and the (x, y) system. We find from the above equations that

$$\begin{bmatrix} x'' \\ y'' \end{bmatrix} = \begin{bmatrix} a'a + b'c & a'b + b'd \\ c'a + d'c & c'b + d'd \end{bmatrix} \begin{bmatrix} x \\ y \end{bmatrix}. \tag{7.5}$$

If we now, without having any thought for the meaning of the exercise, *substitute* for the expression on the left side of (7.3) into (7.4), we obtain

$$\begin{bmatrix} x'' \\ y'' \end{bmatrix} = \begin{bmatrix} a' & b' \\ c' & d' \end{bmatrix} \begin{bmatrix} a & b \\ c & d \end{bmatrix} \begin{bmatrix} x \\ y \end{bmatrix}. \tag{7.6}$$

To give a *meaning* to the latter equation, we need to compare it with (7.5) and *define*

$$\begin{bmatrix} a' & b' \\ c' & d' \end{bmatrix} \begin{bmatrix} a & b \\ c & d \end{bmatrix} = \begin{bmatrix} a'a + b'c & a'b + b'd \\ c'a + d'c & c'b + d'd \end{bmatrix}. \tag{7.7}$$

This shows how we need to *multiply* the two matrices in (7.6) so that the equations (7.6) and (7.5) are equivalent. Thus (7.7) defines matrix multiplication. Note how each of the four elements in the matrix on the right of (7.7) is formed from the elements in one *row* of the first matrix on the left and one *column* of the second matrix on the left. It is worth pausing here to master this, since it makes it easy to remember how we carry out matrix multiplication. We readily see that the operation of multiplication is not commutative. It is not difficult to find matrices that do not commute. For example, we have

$$\begin{bmatrix} 1 & 0 \\ 0 & 0 \end{bmatrix} \begin{bmatrix} 0 & 1 \\ 0 & 0 \end{bmatrix} = \begin{bmatrix} 0 & 1 \\ 0 & 0 \end{bmatrix},$$

whereas, on interchanging the two matrices on the left of the above equation, we obtain

$$\begin{bmatrix} 0 & 1 \\ 0 & 0 \end{bmatrix} \begin{bmatrix} 1 & 0 \\ 0 & 0 \end{bmatrix} = \begin{bmatrix} 0 & 0 \\ 0 & 0 \end{bmatrix}.$$

Although I have mentioned only 2×2 matrices, because they serve my purpose of showing you an algebraic system in which multiplication is not commutative, we can talk about matrices with more than two rows and columns.

Definition 7.1.1 We define the *determinant* of the matrix

$$\mathbf{A} = \begin{bmatrix} a & b \\ c & d \end{bmatrix}$$

as $\det \mathbf{A} = ad - bc$. We say that $\mathbf{A}$ is *nonsingular* if $\det \mathbf{A} \neq 0$. ∎

Let us express (7.7) in the form $\mathbf{AB} = \mathbf{C}$. It is not difficult to verify that

$$\det \mathbf{A} \det \mathbf{B} = \det(\mathbf{AB}). \tag{7.8}$$

It follows from the definition of matrix multiplication given in (7.7) that

$$\mathbf{IA} = \mathbf{AI} = \mathbf{A},$$

where

$$\mathbf{I} = \begin{bmatrix} 1 & 0 \\ 0 & 1 \end{bmatrix}. \tag{7.9}$$

We call $\mathbf{I}$ the *unit matrix*, since it behaves like the number 1 in the algebra of the real numbers. With $\mathbf{A}$ as defined above, we define

$$\lambda \mathbf{A} = \begin{bmatrix} \lambda a & \lambda b \\ \lambda c & \lambda d \end{bmatrix},$$

where λ is any real number. If $\mathbf{A}$ is nonsingular, we define

$$\mathbf{A}^{-1} = \frac{1}{ad - bc} \begin{bmatrix} d & -b \\ -c & a \end{bmatrix}. \tag{7.10}$$

Then we may verify that

$$\mathbf{A}^{-1}\mathbf{A} = \mathbf{A}\mathbf{A}^{-1} = \mathbf{I}.$$

We call the matrix $\mathbf{A}^{-1}$ the *inverse* of the matrix $\mathbf{A}$.

Hamilton introduced the *quaternion*, a quantity involving real numbers a, b, c and d, of the form

$$q = a + bi + cj + dk, \tag{7.11}$$

which reduces to a complex number when $c = d = 0$. Thus $i^2 = -1$, and Hamilton required that, similarly, $j^2 = k^2 = -1$, and that

$$jk = i, \quad ki = j, \quad ij = k, \quad kj = -i, \quad ik = -j, \quad ji = -k. \tag{7.12}$$

We need only write down $jk = i$ and $kj = -i$, and permute i, j, and k in cyclic order, to obtain the other four relations in (7.12). We add and multiply quaternions in an obvious way, and it is clear from (7.12) that the multiplication of quaternions is not commutative. At about the time when Hamilton created his quaternions that are constructed from the four elements 1, i, j, and k, Grassmann defined a system that is constructed from n basic elements. The algebra of Grassmann's system, like Hamilton's, is determined by a multiplication table for the basic elements.

Quaternions can be represented by 2×2 complex matrices. Let us write

$$\begin{bmatrix} w & z \\ -\bar{z} & \bar{w} \end{bmatrix} = \begin{bmatrix} a + bi & c + di \\ -c + di & a - bi \end{bmatrix} = a\mathbf{I} + b\mathbf{H}_i + c\mathbf{H}_j + d\mathbf{H}_k, \tag{7.13}$$

where $\mathbf{I}$ is the unit 2×2 matrix, defined in (7.8), and

$$\mathbf{H}_i = \begin{bmatrix} i & 0 \\ 0 & -i \end{bmatrix}, \quad \mathbf{H}_j = \begin{bmatrix} 0 & 1 \\ -1 & 0 \end{bmatrix}, \quad \mathbf{H}_k = \begin{bmatrix} 0 & i \\ i & 0 \end{bmatrix}. \tag{7.14}$$

There is a one-to-one correspondence between the set of 2×2 matrices defined in (7.13) and the quaternions. The matrix $a\mathbf{I} + b\mathbf{H}_i + c\mathbf{H}_j + d\mathbf{H}_k$ is identified with the quaternion $a + bi + cj + dk$. Note that

$$\mathbf{H}_i^2 = \mathbf{H}_j^2 = \mathbf{H}_k^2 = -\mathbf{I},$$

corresponding to the relations $i^2 = j^2 = k^2 = -1$ given above. We can also verify that

$$\mathbf{H}_j\mathbf{H}_k = \mathbf{H}_i \quad \text{and} \quad \mathbf{H}_k\mathbf{H}_j = -\mathbf{H}_i,$$

and we can permute the suffices i, j, and k cyclically to give the six matrix equations that correspond to the six quaternion equations in (7.12).

Problem 7.1.1 Let

$$\mathbf{A}_1 = \begin{bmatrix} a_1 & -b_1 \\ b_1 & a_1 \end{bmatrix} \quad \text{and} \quad \mathbf{A}_2 = \begin{bmatrix} a_2 & -b_2 \\ b_2 & a_2 \end{bmatrix}.$$

Show that

$$\mathbf{A}_1\mathbf{A}_2 = \begin{bmatrix} a_3 & -b_3 \\ b_3 & a_3 \end{bmatrix} = \mathbf{A}_3,$$

where

$$a_3 = a_1a_2 - b_1b_2 \quad \text{and} \quad b_3 = a_1b_2 + a_2b_1.$$

Show that the relation $\det \mathbf{A}_1 \det \mathbf{A}_2 = \det \mathbf{A}_3$, a special case of (7.8), is equivalent to the identity (1.48). Verify that $\mathbf{A}_1$ and $\mathbf{A}_2$ commute.

Problem 7.1.2 If a_1 and b_1 are not both zero, show that the matrix $\mathbf{A}_1$, defined in Problem 7.1.1, is nonsingular, and show that

$$\mathbf{A}_1^{-1} = \frac{1}{a_1^2 + b_1^2} \begin{bmatrix} a_1 & b_1 \\ -b_1 & a_1 \end{bmatrix}.$$

Problem 7.1.3 As a special case of the matrices defined in Problem 7.1.1, consider the matrix

$$\mathbf{A}_\theta = \begin{bmatrix} \cos\theta & -\sin\theta \\ \sin\theta & \cos\theta \end{bmatrix},$$

for any real θ. Verify that $\mathbf{A}_\theta = \mathbf{I}$ if and only if θ is an integer multiple of 2π. Show that $\mathbf{A}_\theta$ is nonsingular for all values of θ and that

$$\mathbf{A}_\theta^{-1} = \begin{bmatrix} \cos\theta & \sin\theta \\ -\sin\theta & \cos\theta \end{bmatrix}.$$

Verify that $\mathbf{A}_\theta^{-1} = \mathbf{A}_\phi$, where $\phi = -\theta$. Finally, show that

$$\begin{bmatrix} \cos\theta & -\sin\theta \\ \sin\theta & \cos\theta \end{bmatrix} \begin{bmatrix} \cos\phi & -\sin\phi \\ \sin\phi & \cos\phi \end{bmatrix} = \begin{bmatrix} \cos(\theta+\phi) & -\sin(\theta+\phi) \\ \sin(\theta+\phi) & \cos(\theta+\phi) \end{bmatrix}$$

and verify by mathematical induction that

$$\begin{bmatrix} \cos\theta & -\sin\theta \\ \sin\theta & \cos\theta \end{bmatrix}^n = \begin{bmatrix} \cos n\theta & -\sin n\theta \\ \sin n\theta & \cos n\theta \end{bmatrix},$$

where n is a positive integer.

Problem 7.1.4 Show that $ijk = -1$, where i, j, and k are components of the quaternion defined in (7.11).

Problem 7.1.5 Verify that

$$\det \begin{bmatrix} w & z \\ -\bar{z} & \bar{w} \end{bmatrix} = |w|^2 + |z|^2 = a^2 + b^2 + c^2 + d^2,$$

where $w = a + bi$ and $z = c + di$.

Problem 7.1.6 Let $\mathbf{Q} = a\mathbf{I} + b\mathbf{H}_i + c\mathbf{H}_j + d\mathbf{H}_k$, where $\mathbf{H}_i$, $\mathbf{H}_j$, and $\mathbf{H}_k$ are defined in (7.14) and a, b, c, and d are not all zero. Define

$$\mathbf{R} = \frac{1}{\lambda}\left(a\mathbf{I} - b\mathbf{H}_i - c\mathbf{H}_j - d\mathbf{H}_k\right),$$

where $\lambda = a^2 + b^2 + c^2 + d^2$. Verify that $\mathbf{QR} = \mathbf{RQ} = \mathbf{I}$. Equivalently, if $q = a + bi + cj + dk$ and $r = (1/\lambda)(a - bi - cj - dk)$, verify that $qr = rq = 1$.

Problem 7.1.7 Verify that

$$\begin{bmatrix} w_1 & z_1 \\ -\bar{z}_1 & \bar{w}_1 \end{bmatrix} \begin{bmatrix} w_2 & z_2 \\ -\bar{z}_2 & \bar{w}_2 \end{bmatrix} = \begin{bmatrix} w & z \\ -\bar{z} & \bar{w} \end{bmatrix},$$

where

$$w = w_1 w_2 - z_1 \bar{z}_2 \quad \text{and} \quad z = w_1 z_2 + \bar{w}_2 z_1.$$

Problem 7.1.8 If $w_s = a_s + b_s i$ and $z_s = c_s + d_s i$, show that

$$\mathbf{Q}_s = \begin{bmatrix} w_s & z_s \\ -\bar{z}_s & \bar{w}_s \end{bmatrix} = a_s \mathbf{I} + b_s \mathbf{H}_i + c_s \mathbf{H}_j + d_s \mathbf{H}_k.$$

If $\mathbf{Q}_1 \mathbf{Q}_2 = \mathbf{Q} = a\mathbf{I} + b\mathbf{H}_i + c\mathbf{H}_j + d\mathbf{H}_k$, use the result of Problem 7.1.7 to show that

$$a = \mathbf{Re}\,(w_1 w_2 - z_1 \bar{z}_2) = a_1 a_2 - b_1 b_2 - c_1 c_2 - d_1 d_2,$$

where $\mathbf{Re}(z)$ denotes the real part of the complex number z, and show similarly that

$$b = a_1 b_2 + b_1 a_2 + c_1 d_2 - d_1 c_2,$$
$$c = a_1 c_2 - b_1 d_2 + c_1 a_2 + d_1 b_2,$$
$$d = a_1 d_2 + b_1 c_2 - c_1 b_2 + d_1 a_2.$$

Problem 7.1.9 With $\mathbf{Q}_1$, $\mathbf{Q}_2$, and $\mathbf{Q}$ defined as in Problem 7.1.8, use (7.8) to deduce from the relation $\mathbf{Q}_1 \mathbf{Q}_2 = \mathbf{Q}$ that

$$\left(a_1^2 + b_1^2 + c_1^2 + d_1^2\right)\left(a_2^2 + b_2^2 + c_2^2 + d_2^2\right) = a^2 + b^2 + c^2 + d^2, \qquad (7.15)$$

where a, b, c, and d are as given in Problem 7.1.8. Observe that (7.15) generalizes the similar identity involving the moduli of complex numbers, given in (1.64). Note also that (7.15) also generalizes (1.48).

Problem 7.1.10 It is known (see Hardy and Wright [12]) that every prime number can be expressed as a sum of at most four squares. Given this result, deduce from (7.15) that every positive integer can be expressed as a sum of at most four squares.

Problem 7.1.11 Express each of the numbers 17 and 23 as the sum of at most four squares, and hence express $391 = 17 \times 23$ in such a form. In how many ways can you express 391 as a sum of at most four squares?

7.2   Groups

Algebra grew out of a need to manipulate numbers without having to state which particular numbers we are talking about. For example, the relation $a(b+c) = ab+ac$ declares a property shared by all integers, real numbers, complex numbers, quaternions, and indeed the 2×2 matrices that we discussed in the last section. (I did not define addition for matrices. We add two matrices by adding their corresponding elements.) The examples of algebraic systems that have appeared so far in this book have all involved numbers, even those on quaternions and matrices, although the latter two systems also lead us in other directions. However, in this section, we will consider an algebraic system that liberates algebra completely from the bonds of number, and I begin with an illustrative example.

As an aid to understanding what follows you will find it helpful to cut out an equilateral triangle from a sheet of cardboard and label the vertices 1, 2, and 3 on both sides of the triangle-shaped card. Then place the triangle so that the line joining vertices 2 and 3 is horizontal and vertex 1 is at the top.

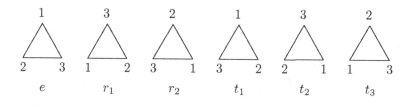

FIGURE 7.1. The group of the equilateral triangle.

The first diagram in Figure 7.1, labeled e, shows the triangle in this position. I have placed the numbers outside the triangle to make them easier to read. Then we can carry out six different operations on the triangle. The first operation is simply to leave it unchanged. The second and third diagrams in Figure 7.1, labeled r_1 and r_2, show the result of rotating the triangle in the first diagram counterclockwise through angles of $\frac{2\pi}{3}$ (one third of a revolution) and $\frac{4\pi}{3}$ (two thirds of a revolution), respectively. The fourth diagram, labeled t_1, shows the result of turning the triangle over by rotating it about its altitude through the top vertex. Similarly, the fifth and sixth diagrams, labeled t_2 and t_3, show the results of turning the triangle over by rotating it about its altitudes through the bottom left and bottom right vertices, respectively. Thus the effect of the first three operations is to leave the triangle face up, and that of the last three operations is to turn it over. Now you see why it is helpful to write the numbers 1, 2, and 3 on both sides of the triangle-shaped card. An alternative way of defining each operation t_j is to say that it is a *reflection* of the triangle, using the appropriate altitude as a mirror.

There are only six permutations of the three objects 1, 2, and 3, and they are all displayed in Figure 7.1. Thus if we carry out any two of the six operations depicted in Figure 7.1 in succession, the triangle must end up in a position that could be achieved by carrying out just one of the six operations. For example, let us perform the operation t_2, defined in the fifth diagram of Figure 7.1, and then perform the operation r_1. Thus we need to rotate the triangle in the fifth diagram counterclockwise through one third of a revolution. This leaves the triangle in the position defined by the fourth diagram in Figure 7.1, which is t_1 We have therefore shown that the operation t_2 followed by r_1 gives t_1, which we will write as

$$t_2 r_1 = t_1.$$

What happens if we carry out these operations in the opposite order, so that we perform the operation r_1 and *then* perform the operation t_2? We find that

$$r_1 t_2 = t_3,$$

showing that $t_2 r_1$ and $r_1 t_2$ are *not* the same. We thus have a system that, like the systems of quaternions and matrices, is not commutative. With a little work, we can find the effect of carrying out any two of the six operations. The results are displayed in Table 7.1, where the first operation is chosen from the column on the left and the second operation is chosen from the row at the top of the table.

$\times$	e	r_1	r_2	t_1	t_2	t_3
e	e	r_1	r_2	t_1	t_2	t_3
r_1	r_1	r_2	e	t_2	t_3	t_1
r_2	r_2	e	r_1	t_3	t_1	t_2
t_1	t_1	t_3	t_2	e	r_2	r_1
t_2	t_2	t_1	t_3	r_1	e	r_2
t_3	t_3	t_2	t_1	r_2	r_1	e

TABLE 7.1. Multiplication table for the group of the equilateral triangle.

Table 7.1 defines a *group*. We refer to e, r_1, r_2, t_1, t_2, and t_3 as the *elements* of the group. The process that combines two elements to produce a third is called a *binary* operation, and we usually refer to this operation as *multiplication*. As we have already seen from $t_2 r_1$ and $r_1 t_2$, the order in which we combine the two elements matters. If x is any one of the six elements, it is clear from the table that

$$ex = xe = x.$$

We refer to e, which behaves like the number 1 in the multiplication of real numbers, as the *identity* element. The identity element appears once in each of the six rows of Table 7.1. We have

$$e^2 = e, \quad r_1 r_2 = r_2 r_1 = e, \quad t_1^2 = t_2^2 = t_3^2 = e,$$

where we have written e^2 as an obvious shorthand for e multiplied by e, as we do in ordinary algebra. We say that e is its own *inverse*, as are t_1, t_2, and t_3, and that r_1 and r_2 are inverses of one another. The group has another property that follows from the way we derived it. If a, b, and c denote any elements of the group, then

$$(ab)c = a(bc). \tag{7.16}$$

To evaluate $(ab)c$ we first multiply a and b (being careful of the order in which we do it, of course), and then multiply ab and c. The element $a(bc)$ on the right of (7.16) is obtained by forming the product bc and *then* multiplying it by a. Any operation of multiplication that satisfies (7.16) is said to be *associative*. Now that we have seen an example of a group, we are able to appreciate a definition that applies to all groups.

Definition 7.2.1 A group is a set of elements G together with a binary operation, which we will call multiplication, that satisfies the following four properties.

1. **Closure** If a and b are elements in G, which need not be distinct, ab is also an element in G.

2. **Associative Law** If a, b, and c are any elements in G, which need not be distinct, then
$$(ab)c = a(bc).$$

3. **Identity** There is an element in G called the identity element, which we will denote by e, such that
$$ea = ae = a \tag{7.17}$$
   for all elements a in G.

4. **Inverse** Corresponding to each element a in G there is an element called its inverse, which we will denote by a^{-1}, such that
$$aa^{-1} = a^{-1}a = e. \tag{7.18}$$

The above four properties are called the group *axioms*. The term "binary operation" mentioned in the above definition is any process in which we combine two elements, taking account of the order in which they are combined, to give a third element. ■

In the English language the words "set" and "group" are synonymous. This is not so in the language of mathematics, where a set is any collection of objects, and a group is a set of objects together with a binary operation that has the properties given in the above definition. Groups with an infinite number of elements are said to be of infinite *order*. If a group G has a finite number of elements, say N, we say that G is of order N.

We have seen from our above example on the group of the equilateral triangle that we can have elements a and b in a group for which $ab \neq ba$, and we say that multiplication in a group is, in general, *noncommutative*. Elements a and b for which $ab = ba$ are said to *commute*. Thus the identity element e commutes with every element in the group, and every element commutes with its inverse. We now state and verify the following results, which are called the *cancellation* properties.

Theorem 7.2.1 (Cancellation Properties) If a, b, and c are any elements in a group and $ac = bc$, then $a = b$. Likewise, if $ca = cb$, then $a = b$.

Proof. If $ac = bc$, then

$$(ac)c^{-1} = (bc)c^{-1}$$

and, using the associative law, we have

$$a(cc^{-1}) = b(cc^{-1}),$$

so that $ae = be$, and thus $a = b$. The second part of the theorem follows similarly. ∎

Example 7.2.1 The quaternion $q = a + bi + cj + dk$ is said to be zero if $a = b = c = d = 0$, and we write $q = 1$ if $a = 1$ and $b = c = d = 0$. We see from Problem 7.1.6 that every nonzero quaternion has an inverse, and that the identity element is $q = 1$. It is clear that the product of two quaternions is a quaternion, and it is easy to see that the associative law holds. Thus the nonzero quaternions form a group. ∎

Example 7.2.2 The set of nonsingular 2×2 matrices with the operation of multiplication forms a group. The identity element is the unit matrix $\mathbf{I}$, which is defined in (7.9), and any nonsingular 2×2 matrix has an inverse, as we saw in (7.10). The closure axiom obviously holds (by the definition of matrix multiplication), and the associative axiom also holds. ∎

Theorem 7.2.2 In any group G there is a unique identity element, and each element in G has a unique inverse.

Proof. Suppose that e_1 and e_2 are both identity elements in G. Since e_1 is an identity element, it follows by writing $e = e_1$ and choosing $a = e_2$ in (7.17) that

$$e_1 e_2 = e_2 e_1 = e_2,$$

and by writing $e = e_2$ and choosing $a = e_1$ in (7.17), we also have

$$e_2 e_1 = e_1 e_2 = e_1.$$

On comparing the latter two equations, we find that $e_1 = e_2$, and so there is a unique identity. Let a denote any element in G, and suppose that x and y are both inverses of a. Thus we have

$$ax = e \qquad \text{and} \qquad ay = e,$$

and we see that $ax = ay$. We deduce from the second part of Theorem 7.2.1 that $x = y$, showing that a has a unique inverse. ∎

Definition 7.2.2 If G is a group in which all elements commute, we say that G is an *abelian* group, named after N. H. Abel (1802–1829). ∎

Example 7.2.3 In this example we use congruences, which are discussed in Section 2.2. Let $G_p = \{1, 2, 3, \ldots, p-1\}$, where p is any prime, and let us combine elements a and b of G_p using multiplication mod p, defining $ab = c$, where c is the unique element of G_p such that

$$ab \equiv c \,(\text{mod } p).$$

We will verify that G_p is an abelian group. Since the commutative property obviously holds, we need only verify that G_p is a group. It follows from how we combine two elements a and b that the closure condition in Definition 7.2.1 is satisfied, and the associative law obviously holds. Next we see that the element 1 is the identity element. It remains only to show that each element has an inverse. Let a, b, and c be any elements in G_p. Then, as we showed in (2.19),

$$ab \equiv ac \,(\text{mod } p) \quad \Rightarrow \quad b \equiv c \,(\text{mod } p),$$

because p divides $ab - ac = a(b - c)$ and p does not divide a. Consequently, p must divide $b - c$, and we deduce that $b = c$. It follows that the $p - 1$ numbers ab, obtained by keeping a fixed and letting b run through all the $p - 1$ elements of G_p, must be all different. In particular, for some choice of b, we must have $ab = 1$, showing that a has an inverse. This completes the proof that G_p is an abelian group. The group multiplication table for G_7 is given in Table 2.5. ∎

For a general group, we have written the result of combining two elements a and b as ab, distinguishing between ab and ba. Since ab looks like the product of a and b in ordinary algebra, it is natural that we describe the binary operation used to combine elements of a group as "multiplication." It should cause no confusion that there are groups in which the set of elements is a subset of the real numbers and the process used to combine elements is *addition*, as in the first of the following two examples.

Example 7.2.4 The set consisting of the integers,

$$\mathbb{Z} = \{0, \pm 1, \pm 2, \pm 3, \dots\},$$

is an abelian group under ordinary addition. This is an infinite group. We combine elements of $\mathbb{Z}$ by using ordinary addition. The closure and associative axioms are obviously satisfied, and the elements commute, because $a + b = b + a$. The identity element is 0, since

$$0 + a = a + 0 = a,$$

for all elements a in $\mathbb{Z}$. The inverse of a is $-a$, since

$$a + (-a) = (-a) + a = 0,$$

for all elements a in $\mathbb{Z}$. ■

	c_1	c_2	c_3	c_4	c_5	c_6
c_1	c_1	c_2	c_3	c_4	c_5	c_6
c_2	c_2	c_1	c_4	c_3	c_6	c_5
c_3	c_3	c_5	c_1	c_6	c_2	c_4
c_4	c_4	c_6	c_2	c_5	c_1	c_3
c_5	c_5	c_3	c_6	c_1	c_4	c_2
c_6	c_6	c_4	c_5	c_2	c_3	c_1

TABLE 7.2. Multiplication table for the cross ratio group.

Example 7.2.5 Let us define $c_j = f_j(t)$, for $1 \leq j \leq 6$, where

$$f_1(t) = t, \qquad f_2(t) = \frac{1}{t}, \qquad f_3(t) = 1 - t,$$

$$f_4(t) = \frac{1}{1-t}, \qquad f_5(t) = \frac{t-1}{t}, \qquad f_6(t) = \frac{t}{t-1}.$$

We also define

$$c_i c_j = f_j(f_i(t)).$$

For example, we find that

$$f_5(f_4(t)) = \frac{f_4(t) - 1}{f_4(t)} = \frac{\frac{1}{1-t} - 1}{\frac{1}{1-t}} = \frac{1 - (1-t)}{1} = t = f_1(t),$$

and thus $c_4 c_5 = c_1$. The complete set of values of $c_i c_j$ is given in Table 7.2. A thorough inspection of this table shows that we have a group with the obvious identity element c_1. It is called the cross ratio group. The elements c_1, c_2, c_3, and c_6 are their own inverses, and the two remaining elements, c_4 and c_5, are inverses of each other. ■

Problem 7.2.1 Let G be any group whose identity element is e, and let a and b be elements of G that satisfy the relation $ab = a$. By using the fact that a has an inverse, show that $b = e$.

Problem 7.2.2 If a and b are elements of a group G, show that the inverse of ab is $b^{-1}a^{-1}$. (This is like saying that the inverse of "Put on socks, put on shoes" is "Take off shoes, take off socks.")

Problem 7.2.3 Consider the set $V = \{e, a, b, c\}$, where $ea = ae = a$, $eb = be = b$, $ec = ce = c$, $e^2 = e$, and

$$a^2 = b^2 = e, \quad ab = ba = c.$$

Verify that this defines the abelian group whose multiplication table is given in Table 7.3. It is called the Klein 4-group, after Felix Klein (1849–1925).

	e	a	b	c
e	e	a	b	c
a	a	e	c	b
b	b	c	e	a
c	c	b	a	e

TABLE 7.3. Multiplication table for the Klein 4-group.

Problem 7.2.4 Make a rectangle $PQRS$ of cardboard and label the vertices on both sides of the card. Let a denote the operation in which the vertices P and Q are interchanged and the vertices S and R are interchanged. Let b denote the operation in which the vertices P and S are interchanged and the vertices Q and R are interchanged. Let c denote the operation of rotating the rectangle through an angle π (half a revolution), and let e denote the operation that leaves the rectangle unchanged. Show that the set of operations $\{e, a, b, c\}$ is a group, and that it is the 4-group.

Problem 7.2.5 Verify that the following table defines an abelian group.

	e	a	b	c
e	e	a	b	c
a	a	b	c	e
b	b	c	e	a
c	c	e	a	b

7.3   Subgroups

In this section we will be concerned with *subsets*. Recall the notation introduced in Definition 2.5.4, where we wrote $S' \subset S$ to denote that S' is a subset of S.

We can see from Table 7.1 that the three elements e, r_1, and r_2 form a group that lies within the group defined by the whole of Table 7.1. This observation prompts the following definition.

Definition 7.3.1 Let G' denote a nonempty subset of a group G. Then we say that G' is a *subgroup* of G if the elements of G' satisfy the four axioms of a group. In particular, this means that the identity element of G must be in G', the inverse of any element of G' must be in G', and the product of any two elements of G' must be in G'.   ∎

Theorem 7.3.1 Let G' denote a nonempty subset of a group G. If the product of any two elements of G' is in G', and the inverse of every element of G' is in G', then G' is a subgroup of G.

Proof. Let a denote an element in G'. Then a^{-1} and $aa^{-1} = e$ belong to G'. Thus G' contains the identity element. The associative axiom holds for G', since it holds for G, and this completes the proof.   ∎

It follows from Definition 7.3.1 that the whole group G is a subgroup of itself, and the set $\{e\}$, consisting only of the identity element, is also a subgroup of G. We regard these as trivial cases and use the term *proper* subgroup to denote a subgroup that is neither $\{e\}$ nor the whole of G. The only proper subgroups of the group defined by Table 7.1 are $\{e, r_1, r_2\}$, $\{e, t_1\}$, $\{e, t_2\}$, and $\{e, t_3\}$.

In Example 7.2.4, we found that the set consisting of the integers,

$$\mathbb{Z} = \{0, \pm 1, \pm 2, \pm 3, \dots\},$$

is an abelian group under ordinary addition. It is easy to verify that all proper subgroups of $\mathbb{Z}$ are given by $\{0, \pm k, \pm 2k, \pm 3k, \dots\}$, where k is an integer greater than 1.

Example 7.3.1 Let us use the group of the equilateral triangle (see Table 7.1), which we will denote by G, to illustrate how to find all subgroups of a given finite group. If r_1 is in a subgroup of G, so is its inverse, which is r_2. Since we have closure for the elements of the set $\{e, r_1, r_2\}$, this is indeed a subgroup. If we try to augment this subgroup by adding t_1, t_2, or t_3, we find that we end up with the whole group G. For example, if we add t_1 to $\{e, r_1, r_2\}$, the closure condition requires us to include $r_1 t_1$, which is t_2, and $r_2 t_1$, which is t_3. So we find that $\{e, r_1, r_2\}$ is the only proper subgroup of G that includes r_1 or r_2. If we begin with t_1, we need to include its inverse, which is t_1 itself. Since we have closure for the set $\{e, t_1\}$, this is a proper subgroup of G, and we see that the same is true of $\{e, t_2\}$, and $\{e, t_3\}$. If we try to add any another element to $\{e, t_1\}$, the closure condition forces us to include all the elements of G, and the same is true if we try to add any another element to $\{e, t_2\}$ or $\{e, t_3\}$. Thus G has the four proper subgroups $\{e, t_1\}$, $\{e, t_2\}$, $\{e, t_3\}$, and $\{e, r_1, r_2\}$. ∎

Example 7.3.2 Consider $S = \{\pm 1, \pm i, \pm j, \pm k\}$, a subset of the quaternions. We see from (7.12) that products of elements of S belong to S and that the inverses of all elements of S are in S. Thus, by Theorem 7.3.1, S is a subgroup of the quaternions. ∎

Example 7.3.3 Let G denote the group of 2×2 nonsingular matrices and let $G' \subset G$ denote the set of matrices of the form

$$\mathbf{A} = \begin{bmatrix} a & -b \\ b & a \end{bmatrix},$$

where a and b are not both zero, so that $\mathbf{A}$ is nonsingular. Then we see from the results in Problems 7.1.1 and 7.1.2 that the product of any two elements of G' is in G', and the inverse of every element of G' is in G'. Thus it follows from Theorem 7.3.1 that G' is a subgroup of G. ∎

Let a denote any element of a group G of order n whose identity element is e, and let us consider the infinite sequence

$$e, a, a^2, a^3, \dots . \tag{7.19}$$

It is convenient to define $a^0 = e$ and $a^1 = a$. This is consistent with the behavior of the identity element in group multiplication, since the property $ea = ae = a$, defined in (7.17), corresponds to

$$a^0 a^1 = a^1 a^0 = a^1,$$

a special case of the "law of indices"

$$a^j a^k = a^{j+k}$$

in ordinary algebra. The elements in the infinite sequence (7.19) cannot all be different, since G is finite. We now state and prove a helpful result.

Theorem 7.3.2 Let a be an element of a group G. If $a^j = a^k$, where $j > k \geq 0$, then $a^{j-k} = e$.

Proof. If $k = 0$, there is nothing to prove. If $k \geq 1$, we can rewrite $a^j = a^k$ as

$$a^{j-1}a = a^{k-1}a,$$

and we then obtain from the cancellation property in Theorem 7.2.1 that

$$a^{j-1} = a^{k-1}.$$

We can continue to reduce j and k by 1 until we obtain $a^{j-k} = e$. This completes the proof. ∎

We conclude that in any finite group, some power of every element is equal to the group's identity element. In the following definition it should cause no confusion that we have already used the word "order" to denote the number of elements in a finite group.

Definition 7.3.2 Let G be a finite group with identity element e. For any element a, let m denote the smallest integer such that $a^m = e$. We call m the order of a. ∎

It follows directly from Definition 7.3.2 that the identity element in a group is the only element of order 1. If a has order m, the elements of

$$S = \{e, a, a^2, a^3, \ldots, a^{m-1}\} \tag{7.20}$$

are all different. For if $a^j = a^k$ with $0 \le k < j \le m - 1$, then it would follow from Theorem 7.3.2 that $a^{j-k} = e$, where $0 < j - k \le m - 1$, and this would contradict the statement that the order of a is m. We also note that

$$a^{qm} = (a^m)^q = e^q = e, \tag{7.21}$$

where q is any positive integer. We can show that for any choice of element a in the group G, the set S in (7.20) is a subgroup of G. To verify the closure axiom for S, we have

$$a^i a^j = a^{i+j},$$

where both i and j lie between 0 and $m-1$, and we see that a^{i+j} is obviously in S if $i + j \le m$. If $i + j > m$, with $0 \le i, j \le m - 1$, we can write

$$a^i a^j = a^{i+j} = a^{i+j-m} a^m = a^{i+j-m} e = a^{i+j-m},$$

and since in this case $0 < i + j - m \le m - 2$, we see that a^{i+j-m} is in S. Then we observe from

$$a^j a^{m-j} = a^m = e \quad \text{for} \quad 1 \le j \le m - 1$$

that a^{m-j}, which is in S, is the inverse of a^j. It follows from Theorem 7.3.1 that S is a subgroup of G. The group S is called the *cyclic* group of order m, which we will henceforth denote by C_m.

For example, the group C_6 is the set $\{1, z, z^2, z^3, z^4, z^5\}$, where the binary operation is the multiplication of complex arithmetic and $z = \frac{1}{2} + \frac{\sqrt{3}}{2} i$. (See Problem 1.5.4.) We will see later, when we discuss isomorphisms, that a group can appear in different guises, and this is not the only way of expressing the group C_6.

Theorem 7.3.3 Let a be an element of a group G. If a is of order m, then $a^k = e$ if and only if k is a multiple of m.

Proof. Let us apply the division algorithm (see Definition 2.1.1) and write

$$k = qm + r, \tag{7.22}$$

where the quotient q is a nonnegative integer and the remainder r satisfies the inequalities $0 \leq r < m$. Note that $q = 0$ and $r = k$ when $k < m$. Then we derive from (7.22) and (7.21) that

$$a^k = a^{qm+r} = a^{qm} a^r = ea^r = a^r.$$

Since $0 \leq r < m$, $a^k = e$ if and only if $r = 0$, and thus $k = qm$. ■

Example 7.3.4 It is easily verified from Table 7.1 that for the group of the equilateral triangle, the identity element has order 1 (as always), r_1 and r_2 have order 3, and the remaining elements, t_1, t_2, and t_3, all have order 2. ■

Example 7.3.5 In Example 7.2.3 we discussed the group G_p of residues modulo p, where p is any prime. Let us now consider the group G_{13}. We can verify that, for example,

$$3^2 \equiv 9 \,(\text{mod } 13) \quad \text{and} \quad 3^3 \equiv 1 \,(\text{mod } 13),$$

so that the element 3 in G_{13} has order 3. With a little work, we find that the element 12 has order 2, the elements 3 and 9 have order 3, the elements 5 and 8 have order 4, the elements 4 and 10 have order 6, and the elements 2, 6, 7, and 11 have order 12. ■

We use the word *premultiply* to denote multiplication on the left, and *postmultiply* to denote multiplication on the right. Thus ab can be described as b premultiplied by a, or as a postmultiplied by b.

Definition 7.3.3 Let

$$H = \{h_1, h_2, \ldots\} \tag{7.23}$$

denote a subgroup of a group G, and let $a \in G$. Then we define

$$aH = \{ah_1, ah_2, \ldots\} \tag{7.24}$$

and call aH a *left coset* of G relative to the subgroup H. We similarly define

$$Ha = \{h_1 a, h_2 a, \ldots\}, \tag{7.25}$$

and call Ha a *right coset* of G relative to the subgroup H. ■

It is easy to verify that the elements of the coset aH are all different, as are the elements of the coset Ha. For if $ah_i = ah_j$ or $h_i a = h_j a$, we can use the cancellation properties (see Theorem 7.2.1) to show that this can happen only when $h_i = h_j$. In the above definition we have chosen a to be any element of the group G. In the special case in which a is itself in the subgroup H, then $aH = Ha = H$. (See Problem 7.3.5.)

Example 7.3.6 We have already noted that $H = \{e, r_1, r_2\}$ is a subgroup of the group of the equilateral triangle, whose multiplication table is given in Table 7.1. Let us evaluate the left cosets $t_1 H$ and $t_2 H$. We find that

$$t_1 H = \{t_1 e, t_1 r_1, t_1 r_2\} = \{t_1, t_3, t_2\}$$

and

$$t_2 H = \{t_2 e, t_2 r_1, t_2 r_2\} = \{t_2, t_1, t_3\}.$$

Since the order in which we write down the elements of a set does not matter, we see that the two cosets $t_1 H$ and $t_2 H$ are the same, and you should verify that they are also equal to the coset $t_3 H$.   ∎

Example 7.3.6 demonstrates that when a is not an element of the subgroup H, the coset aH is not necessarily a *subgroup* of the group G. We can now explore how the concept of a coset helps us verify one of the fundamental properties of subgroups of a finite group. This is Lagrange's theorem, named after J. L. Lagrange (1736–1813).

Theorem 7.3.4 (Lagrange's theorem) Given any finite group G, the order of any subgroup of G is a divisor of the order of G.

Proof. Let the group G have n elements, and let H be a subgroup of G with elements $h_1, h_2, \ldots, h_m$. We will show that m is a divisor of n.

If $m = 1$ or if H is the whole of G, so that $m = n$, there is nothing to prove. Let us therefore assume that H is a proper subgroup of G, so that $1 < m < n$, and choose an element a_1 that is in G but is not in H. Then the elements of the coset $a_1 H$ give m different elements of the group G. Let a_2 denote an element of G that is not in the coset $a_1 H$, and let us consider the coset $a_2 H$, whose m elements are all different. We have reached the key step in the proof, which is to show that the cosets $a_1 H$ and $a_2 H$, whose elements all belong to G, have no element in common. For otherwise we would have

$$a_1 h_i = a_2 h_j,$$

for some values of i and j. If we postmultiply (remember that this means multiply on the right) both sides of the above equation by h_j^{-1}, we find that

$$a_1 h_i h_j^{-1} = a_2.$$

Since H is a subgroup, $h_i h_j^{-1}$ is an element of H and so the last equation shows that a_2 would have to be an element of the coset $a_1 H$, contrary to how we chose it! This shows that the two cosets give $2m$ different elements of the group G. If there is an element a_3 in G that is not in $a_1 H$ or $a_2 H$, we can construct the coset $a_3 H$, and a similar argument to the one we have just used shows that the cosets $a_1 H$, $a_2 H$, and $a_3 H$ give $3m$ different elements of G. We continue with this construction, and since G is finite, this process must end after, say, k steps. We thus find that each element of

G belongs to precisely one of the k cosets $a_1H, a_2H, \ldots, a_kH$, and hence $n = mk$. This completes the proof. ∎

The next two theorems follow easily from Lagrange's theorem.

Theorem 7.3.5 Let G denote a group of order n and let a denote an element of G that is of order m. Then m is a divisor of n.

Proof. We saw above that the set

$$C_m = \{e, a, a^2, a^3, \ldots, a^{m-1}\},$$

originally denoted by S in (7.20), is a subgroup of G. We note that C_m is of order m, and it follows from Lagrange's theorem, Theorem 7.3.4 that m is a divisor of n. This completes the proof. ∎

Theorem 7.3.6 If a group G is of order p, where p is a prime number, then G is the cyclic group C_p.

Proof. It follows from Theorem 7.3.5 that the order of any element of G must be either 1 or p. If a is any element of G not equal to e, then a must be of order p. The p elements $e, a, a^2, \ldots, a^{p-1}$ are all distinct and all belong to G, and so they must be the elements of G. Thus G is the cyclic group C_p. ∎

Definition 7.3.4 Let a be an element of an arbitrary group G, and let H_a denote the set of all elements of G that commute with a. Thus $h \in H_a$ if and only if $ah = ha$. We call H_a the *normalizer* of a. ∎

Theorem 7.3.7 Let a be any element of an arbitrary group G. Then H_a, the normalizer of a, is a subgroup of G.

Proof. First we note that H_a contains the identity element, and we see from Theorem 7.3.1 that we need only show that H satisfies the closure and inverse axioms of a group. If h_1 and h_2 are both in H_a, we have

$$a(h_1h_2) = (ah_1)h_2 = (h_1a)h_2 = h_1(ah_2) = h_1(h_2a) = (h_1h_2)a. \quad (7.26)$$

Note that to verify the chain of equalities in (7.26), we need to use both the associative axiom in Definition 7.2.1 and the fact that h_1 and h_2 belong to H_a. We see from (7.26) that $a(h_1h_2) = (h_1h_2)a$, showing that the closure axiom for H_a is satisfied.

Let h be any element in the set H_a, so that $ah = ha$. If we premultiply the element on each side of the latter equation by h^{-1} and also postmultiply by h^{-1}, we obtain $h^{-1}a = ah^{-1}$, showing that the inverse of h is also in H_a. This completes the proof. ∎

Example 7.3.7 Consider the group defined by Table 7.1. We find that the normalizer of the element r_1 is the subgroup $\{e, r_1, r_2\}$, and that the normalizer of r_2 is also $\{e, r_1, r_2\}$. I leave it to the reader to find the normalizers of t_1, t_2, and t_3. ∎

Definition 7.3.5 Let H denote a subgroup of a group G such that the left coset aH is equal to the right coset Ha for every $a \in G$. Then we say that H is a *normal* subgroup of G. ∎

Thus H is a normal subgroup if every left coset is also a right coset. If G is abelian, then all its subgroups will be normal, including G itself. The concept of normal subgroup is of interest only when G is not abelian.

Definition 7.3.6 A group is called *simple* if it has no proper normal subgroups. ∎

Problem 7.3.1 Consider the group G' consisting of matrices of the form

$$\mathbf{A} = \begin{bmatrix} a & -b \\ b & a \end{bmatrix},$$

where the real numbers a and b are not both zero, which we discussed in Example 7.3.3. Let G'' denote the set of those matrices in G' for which $a^2 + b^2 = 1$. Verify that G'' is a subgroup of G' .

Problem 7.3.2 Find all the proper subgroups of the group defined in Example 7.2.5.

Problem 7.3.3 Use the group multiplication table given in Example 7.2.5 to determine the order of every element in this group.

Problem 7.3.4 Let a be an element of a finite group G. Prove that a^{-1} is a power of a.

Problem 7.3.5 Let H be a subgroup of a group G. Consider the left coset aH for the special case $a \in H$. Deduce from the closure axiom that $aH \subset H$. Let h denote any element of H. Argue that since $a^{-1} \in H$, we have $a^{-1}h \in H$, and so $h = a(a^{-1}h) \in H$, showing that $H \subset aH$. Deduce from $aH \subset H$ and $H \subset aH$ that $aH = H$.

Problem 7.3.6 Let the element a in a group G have order m. Deduce from

$$a^m(a^{-1})^m = (aa \cdots a)(a^{-1}a^{-1} \cdots a^{-1}) = e$$

that $(a^{-1})^m = e$, so that the order of a^{-1} is m or less. Suppose that the order of a^{-1} is $k < m$, and deduce from $a^k(a^{-1})^k = e$ that $a^k = e$, contradicting the above statement that a is of order m and thus showing that a^{-1} has the same order as a.

7.4   Permutation Groups

In Section 5.1 we saw that there are $n!$ arrangements or permutations of n objects. Let us name the objects $1, 2, \ldots, n$. We will write

$$\begin{pmatrix} 1 & 2 & 3 & 4 \\ 3 & 1 & 4 & 2 \end{pmatrix}$$

to denote the permutation in which, as we say, 1 is mapped to 3, 2 is mapped to 1, 3 is mapped to 4, and 4 is mapped to 2. Observe that this permutation can be expressed in $4! = 24$ ways, one other way being

$$\begin{pmatrix} 2 & 4 & 3 & 1 \\ 1 & 2 & 4 & 3 \end{pmatrix}.$$

More generally, we write

$$\begin{pmatrix} 1 & 2 & \cdots & n \\ m_1 & m_2 & \cdots & m_n \end{pmatrix}$$

to denote the permutation in which 1 is mapped to m_1, 2 is mapped to m_2, and so on, where $m_1, m_2, \ldots, m_n$ is a permutation of the first n positive integers. For example, let

$$a_1 = \begin{pmatrix} 1 & 2 & 3 & 4 \\ 3 & 1 & 4 & 2 \end{pmatrix} \quad \text{and} \quad a_2 = \begin{pmatrix} 1 & 2 & 3 & 4 \\ 2 & 3 & 4 & 1 \end{pmatrix}.$$

Then we define

$$a_1 a_2 = \begin{pmatrix} 1 & 2 & 3 & 4 \\ 3 & 1 & 4 & 2 \end{pmatrix} \begin{pmatrix} 1 & 2 & 3 & 4 \\ 2 & 3 & 4 & 1 \end{pmatrix} = \begin{pmatrix} 1 & 2 & 3 & 4 \\ 4 & 2 & 1 & 3 \end{pmatrix}.$$

In the line above, $a_1 a_2$ denotes the permutation that results from first carrying out the permutation a_1 and then carrying out the permutation a_2. In carrying out the permutation a_1, 1 is mapped to 3, and then 3 is mapped to 4 by the permutation a_2. Thus the total effect of $a_1 a_2$ is to map 1 to 4. Similarly, a_1 maps 2 to 1 and a_2 maps 1 to 2, and so the effect of $a_1 a_2$ is to map 2 to itself. You should check that the permutation $a_1 a_2$ maps 3 to 1 and 4 to 3.

Theorem 7.4.1 The $n!$ permutations of n objects form a group.

Proof. We can see that the closure axiom (see Definition 7.2.1) is satisfied from the way we have defined $a_1 a_2$ above, and it is also clear that

$$(a_1 a_2) a_3 = a_1 (a_2 a_3),$$

showing that the associative axiom holds. There is an identity element,

$$e = \begin{pmatrix} 1 & 2 & \cdots & n \\ 1 & 2 & \cdots & n \end{pmatrix}.$$

Finally, if

$$a_1 = \begin{pmatrix} 1 & 2 & \cdots & n \\ m_1 & m_2 & \cdots & m_n \end{pmatrix},$$

then we can define

$$a_1^{-1} = \begin{pmatrix} m_1 & m_2 & \cdots & m_n \\ 1 & 2 & \cdots & n \end{pmatrix},$$

and we see that

$$a_1 a_1^{-1} = \begin{pmatrix} 1 & 2 & \cdots & n \\ m_1 & m_2 & \cdots & m_n \end{pmatrix} \begin{pmatrix} m_1 & m_2 & \cdots & m_n \\ 1 & 2 & \cdots & n \end{pmatrix},$$

so that

$$a_1 a_1^{-1} = \begin{pmatrix} 1 & 2 & \cdots & n \\ 1 & 2 & \cdots & n \end{pmatrix} = e.$$

Similarly, we can show that $a_1^{-1} a_1 = e$, and this completes the proof that the set of $n!$ permutations of n objects forms a group. It called the *symmetric group*, and we will denote it by S_n. ■

Definition 7.4.1 A *cycle* of period n is a permutation of the form

$$c = \begin{pmatrix} a_1 & a_2 & \cdots & a_{n-1} & a_n \\ a_2 & a_3 & \cdots & a_n & a_1 \end{pmatrix}.$$

We will usually express the above cycle in the shorter form

$$c = (a_1, a_2, \ldots, a_n). ■$$

Example 7.4.1 Consider the permutation

$$\begin{pmatrix} 1 & 2 & 3 & 4 & 5 & 6 & 7 & 8 \\ 6 & 1 & 8 & 3 & 5 & 7 & 2 & 4 \end{pmatrix},$$

which belongs to S_8. We see that this permutation maps 1 to 6, maps 6 to 7, maps 7 to 2, and maps 2 to 1, giving the cycle $c_1 = (1, 6, 7, 2)$. Let us find a number that is not in the cycle c_1, for example, 3. The above permutation maps 3 to 8, maps 8 to 4, and maps 4 to 3, giving the cycle $c_2 = (3, 8, 4)$. The only number that is not in c_1 or c_2 is 5, which is mapped to itself, giving the trivial cycle $c_3 = (5)$. Note that from the way they were constructed, the cycles c_1, c_2, and c_3 operate on different subsets of the set $\{1, 2, \ldots, n\}$. We say that the cycles c_1, c_2, and c_3 are *disjoint*, and it can be proved that

$$\begin{pmatrix} 1 & 2 & 3 & 4 & 5 & 6 & 7 & 8 \\ 6 & 1 & 8 & 3 & 5 & 7 & 2 & 4 \end{pmatrix} = c_1 c_2 c_3 = (1, 6, 7, 2)(3, 8, 4)(5),$$

so that the above permutation is expressed as a product of three disjoint cycles. ■

Example 7.4.1 illustrates the following general result, which I state without giving a formal proof.

Theorem 7.4.2 Every permutation in S_n can be expressed uniquely as a product of one or more disjoint cycles.   ■

Let us write

$$p = c_1 c_2 \cdots c_k,$$

where $p \in S_n$ is a permutation expressed as a product of the disjoint cycles $c_1, c_2, \ldots, c_k$. Since these cycles are disjoint, they commute, and so

$$p^2 = (c_1 c_2 \cdots c_k)(c_1 c_2 \cdots c_k) = c_1^2 c_2^2 \cdots c_k^2.$$

We can see that more generally,

$$p^j = c_1^j c_2^j \cdots c_k^j,$$

for any positive integer j.

Example 7.4.2 Consider the permutation group S_3, which consists of the six elements of the form

$$\begin{pmatrix} 1 & 2 & 3 \\ m_1 & m_2 & m_3 \end{pmatrix},$$

where m_1, m_2, and m_3 are the numbers 1, 2, and 3 in some order. We have already met this group (also called the group of the equilateral triangle) at the beginning of this chapter, and we named its elements e, r_1, r_2, t_1, t_2, and t_3. We know from Theorem 7.3.5 that the order of each element must be a divisor of 6. Therefore, each element other than the identity element must be of order 2 or 3. Let us express all six elements as products of disjoint cycles. You should check that these are

$$e = (1)(2)(3), \quad r_1 = (1,3,2), \quad r_2 = (1,2,3),$$
$$t_1 = (1)(2,3), \quad t_2 = (2)(1,3), \quad t_3 = (3)(1,2).$$

We note that

$$r_1^3 = r_2^3 = e \quad \text{and} \quad t_1^2 = t_2^2 = t_3^2 = e, \tag{7.27}$$

so that r_1 and r_2 have order 3, while t_1, t_2, and t_3 have order 2. Since, for example, t_1 and t_2 do not commute, this group is not abelian. It has four proper subgroups, as already noted, namely $\{e, r_1, r_2\}$, $\{e, t_1\}$, $\{e, t_2\}$, and $\{e, t_3\}$. It can be verified that only the first of these, $\{e, r_1, r_2\}$, is a normal subgroup.   ■

Definition 7.4.2 A *transposition* is a cycle of period 2.   ■

Definition 7.4.3 A permutation is called *even* if it can be obtained by making an even number (including zero) of transpositions, and is otherwise called *odd*.   ■

Definition 7.4.4 We can show that A_n, defined as the set of all even permutations of n objects, is a group. It is a subgroup of the full permutation group S_n and is called the *alternating group*.   ∎

We can show that the alternating group A_n has $\frac{1}{2}n!$ elements, half as many as S_n.

Problem 7.4.1 Express the permutation

$$p = \begin{pmatrix} 1 & 2 & 3 & 4 & 5 & 6 & 7 & 8 & 9 \\ 8 & 9 & 6 & 1 & 3 & 5 & 4 & 7 & 2 \end{pmatrix}$$

as a product of disjoint cycles. Use this to evaluate the permutation p^3.

Problem 7.4.2 Consider the decomposition of all permutations of the form

$$\begin{pmatrix} 1 & 2 & 3 & 4 \\ m_1 & m_2 & m_3 & m_4 \end{pmatrix}$$

into disjoint cycles, for example,

$$\begin{pmatrix} 1 & 2 & 3 & 4 \\ 4 & 3 & 2 & 1 \end{pmatrix} = (1,4)(2,3).$$

Show that there are three permutations that can be expressed in the form $(m_1, m_2)(m_3, m_4)$, six permutations that can be expressed in the form (m_1, m_2, m_3, m_4), eight in the form $(m_1, m_2, m_3)(m_4)$, six in the form $(m_1, m_2)(m_3)(m_4)$, and one in the form $(m_1)(m_2)(m_3)(m_4)$, the latter being the identity element. Deduce that the permutation group of order four consists of six elements of order 4, eight elements of order 3, nine elements of order 2, and the identity element.

Problem 7.4.3 Given the cycle $c_1 = (1, 6, 7, 2)$, which we encountered in Example 7.4.1, show that

$$c_1^2 = (1, 7)(2, 6) \quad \text{and} \quad c_1^3 = (1, 2, 7, 6),$$

and hence evaluate c_1^j for all $j \geq 1$.

7.5   Further Material

Consider the set $\{1, -1, i, -i\}$, where $i^2 = -1$. If we combine any two members of this set, using the multiplication of complex arithmetic, we find that the set forms form a group whose multiplication table is given in Table 7.4. The identity element is 1. We observe that the element i has

	1	−1	i	$-i$
1	1	−1	i	$-i$
−1	−1	1	$-i$	i
i	i	$-i$	−1	1
$-i$	$-i$	i	1	−1

TABLE 7.4. Multiplication table for the group with elements $1, -1, i, -i$.

order 4. Thus all elements of the group are generated by the powers of i,

$$i^0 = 1, \quad i^1 = i, \quad i^2 = -1, \quad i^3 = -i. \tag{7.28}$$

Let us look again at the group discussed in Problem 7.2.5, whose elements are e, a, b, and c. We see from the group multiplication table that

$$a^0 = e, \quad a^1 = a, \quad a^2 = b, \quad a^3 = c. \tag{7.29}$$

Compare (7.28) and (7.29). If we identify 1 with e, i with a, -1 with b, and $-i$ with c, the two group multiplication tables are the same. We say that the two groups are *isomorphic* and regard them as being the same. If we let $\mathbf{I}$ denote the 2×2 unit matrix and let $\mathbf{H}$ denote any of the 2×2 matrices $\mathbf{H}_i$, $\mathbf{H}_j$, and $\mathbf{H}_k$ defined in (7.14), then $\mathbf{H}^2 = -\mathbf{I}$, and we can verify that $\{\mathbf{I}, -\mathbf{I}, \mathbf{H}, -\mathbf{H}\}$ forms a group that is isomorphic to the above two groups through the identification of $\mathbf{H}$ with i and a.

Definition 7.5.1 Two groups $G = \{a, b, \dots\}$ and $G' = \{a', b', \dots\}$ are said to be *isomorphic* if there is a one-to-one correspondence (see Definition 2.5.1)

$$a \leftrightarrow a', \ b \leftrightarrow b', \ \dots$$

between their elements such that

$$ab = c \quad \text{implies} \quad a'b' = c',$$

and vice versa.   ∎

When we encounter a group it is helpful to establish whether it is isomorphic to any group with which we are already familiar. Obviously, for two *finite* groups to be isomorphic they need to be of the same order and must contain the same number of elements of a given order. It is usually not enough (at least for ordinary mortals) to glance at the multiplication tables of the two groups. For the elements may be labeled differently and will not necessarily appear in the same order.

Example 7.5.1 Let us consider again the group that we discussed in Example 7.2.5. We identified c_1 as the identity element of this group, and so it has order 1. For the sake of clarity, let us write c_1 as e. We will now determine the order of all the other elements. We find that

$$c_2^2 = c_3^2 = c_6^2 = e \quad \text{and} \quad c_4^3 = c_5^3 = e.$$

If we look back at Example 7.4.2, we see from (7.27) that we have already studied a group of order six in which three elements have order 2 and two elements have order 3. Then a further inspection of the group multiplication tables reveals that we can make the correspondence

$$r_1 \leftrightarrow c_5, \quad r_2 \leftrightarrow c_4,$$

and

$$t_1 \leftrightarrow c_2, \quad t_2 \leftrightarrow c_3, \quad t_3 \leftrightarrow c_6,$$

and see that the two groups are indeed isomorphic. The one-to-one correspondence is not unique.   ∎

We can construct more groups by combining two groups in a simple way, by taking their *direct product*, which we now define.

Definition 7.5.2 Let G and H be groups of orders m and n, respectively. We will write $G \times H$ to denote the set of mn ordered pairs of the form (g, h), where g belongs to G and h belongs to H. We combine elements of $G \times H$, defining the product of (g_1, h_1) and (g_2, h_2) as

$$(g_1, h_1)(g_2, h_2) = (g_1 g_2, h_1 h_2). \tag{7.30}$$

Note that on the right side of (7.30), $g_1 g_2$ is an element of G and $h_1 h_2$ is an element of H. Thus (7.30) shows how two elements of $G \times H$ are combined to give an element of $G \times H$. We call $G \times H$ the *direct product* of the groups G and H.   ∎

We will now show that with the binary operation defined by (7.30), the set $G \times H$ is a group.

Theorem 7.5.1 Given two groups G and H of orders m and n, respectively, their direct product $G \times H$ is a group of order mn. Further, if G and H are both abelian, their direct product is also abelian.

Proof. The closure axiom for $G \times H$ follows immediately from (7.30), and we also see from (7.30) that the associative property holds for $G \times H$ because it holds for the groups G and H. If e_G and e_H denote the identity elements for G and H, respectively, we see from (7.30) that (e_G, e_H) is the identity element for $G \times H$. Finally, we see that

$$(g, h)(g^{-1}, h^{-1}) = (g^{-1}, h^{-1})(g, h) = (e_G, e_H),$$

and thus (g^{-1}, h^{-1}) is the inverse of (g, h). This shows that $G \times H$ is indeed a group. If G and H are both abelian,

$$(g_1, h_1)(g_2, h_2) = (g_1 g_2, h_1 h_2) = (g_2 g_1, h_2 h_1) = (g_2, h_2)(g_1, h_1)$$

and thus their direct product $G \times H$ is also abelian.   ∎

Example 7.5.2 Let $G = \{e, g\}$, where e is the identity element of G and $g^2 = e$. Thus $G = C_2$, the cyclic group of order two, and the group $G \times G$ has four elements, which we will write as

$$e^* = (e, e), \quad p = (e, g), \quad q = (g, e), \quad r = (g, g).$$

Thus e^* is the identity element of $G \times G$, and

$$p^2 = q^2 = e^*, \quad pq = qp = r.$$

We see that $G \times G$ is the 4-group, which we met in Problem 7.2.3. ■

We can use a simpler notation for working with direct products. We can write gh instead of (g, h), where g is an element of the group G and h is an element of the group H, on the understanding that every element of G commutes with every element of H. Thus the relation defining the product of two elements,

$$(g_1, h_1)(g_2, h_2) = (g_3, h_3),$$

where where $g_3 = g_1 g_2$ is an element of G and $h_3 = h_1 h_2$ is an element of H, can be expressed as

$$g_1 h_1 g_2 h_2 = g_1 g_2 h_1 h_2 = g_3 h_3.$$

See Problem 7.5.2 for a formal justification of this simpler notation for handling direct products. To show the advantages of the simpler notation, let us rework Example 7.5.2.

Example 7.5.3 Let us write

$$G_a = \{e, a\} \quad \text{and} \quad G_b = \{e, b\}, \quad \text{where} \quad a^2 = b^2 = e,$$

and e is the identity element common to G_a and G_b, which are both isomorphic to the cyclic group C_2. Then $C_2 \times C_2$ is isomorphic to

$$G_a \times G_b = \{e, a, b, ab\}, \quad \text{where} \quad a^2 = b^2 = e \quad \text{and} \quad ab = ba,$$

which, as expected from Example 7.5.2, is the 4-group. Note that since we are taking the direct product of C_2 with itself, we need to distinguish between the two *copies* of C_2. This is the reason for using the two isomorphic groups G_a and G_b. ■

Suppose we have found $G_1 \times G_2$, the direct product of the groups G_1 and G_2. We can then find the direct product of $G_1 \times G_2$ with a third group G_3, to give the group $(G_1 \times G_2) \times G_3$. Since the latter group is isomorphic to $G_1 \times (G_2 \times G_3)$, we can denote it more simply by $G_1 \times G_2 \times G_3$. (I will omit a justification of this isomorphism.)

Let $S = \{a_1, a_2, a_3, \ldots, a_m\}$ denote a subset of the elements of a given finite group G. Let S' denote the set of all the elements of G that are obtained by taking all possible products of a finite number of elements chosen from S, where we allow repetitions of the same element, for example $a_2^2 a_3 a_1 a_2$. (In fact, it can be proved that the set S' is a *subgroup* of G, and Problem 7.3.4 provides part of the proof.) Since G is finite, we can, if necessary, add further elements of G to S until the corresponding set S' becomes equal to G. We then say that S is a set of *generators* of the group G. We can always find such a set S. For example, we could begin by choosing $S = G$. Now suppose that S is a set of generators of the group G and that there is an element a in S that can be expressed in terms of the other elements of S. Then we can remove a from S, and the resulting set will still generate G.

Example 7.5.4 Consider the group of the equilateral triangle, defined by Table 7.1, which we found to be isomorphic to S_3, the permutation group of order 3. One set of generators for S_3 is $\{e, r_1, t_1\}$. You should check that each of the sets $\{e, r_i, t_j\}$, where $i = 1$ or 2 and $j = 1$, 2, or 3, is a set of generators for S_3.

   The 4-group, defined in Problem 7.2.3, is generated by the set that consists of e and any two of the other three elements.   ■

Theorem 7.5.2 If every element of a group G, other than the identity element, is of order 2, then G is abelian and is isomorphic to

$$C_2 \times C_2 \times \cdots \times C_2,$$

where C_2 is the cyclic group of order 2. Consequently, the order of the group G is a power of 2.

Proof. This result is obviously true when G is of order 2, since G must then be isomorphic to C_2. Suppose that G has order greater than 2. Let a_1 and a_2 be any two distinct elements of G other than the identity e. Then, since

$$a_1^2 = a_2^2 = e,$$

we deduce that

$$a_1 = a_1^{-1} \quad \text{and} \quad a_2 = a_2^{-1}.$$

By the closure axiom, $a_1 a_2$ is an element of G. We see that $a_1 a_2 \neq e$, for otherwise,

$$a_1 = a_2^{-1} = a_2.$$

It is also clear from the cancellation property (see Theorem 7.2.1) that $a_1 a_2$ is distinct from both a_1 and a_2. Since the element $a_1 a_2$ has order 2, we have

$$e = (a_1 a_2)^2 = (a_1 a_2)(a_1 a_2),$$

and thus

$$e = a_1(a_2 a_1)a_2.$$

If we now premultiply the elements on each side of the latter equation by a_1, postmultiply by a_2, and use the properties that $a_1^2 = a_2^2 = e$, we obtain

$$a_1 a_2 = a_2 a_1.$$

Since every pair of elements of the group G commute, it is abelian. Let $S = \{a_1, a_2, \ldots, a_m\}$ denote a set of independent generators for G. Since G is abelian and each element in S, being an element of G, is of order 2, we can write any element of G in the form

$$a_1^{\alpha_1} a_2^{\alpha_2} \cdots a_m^{\alpha_m},$$

where each exponent α_j has the value 0 or 1. It follows that

$$G = \{e, a_1\} \times \{e, a_2\} \times \cdots \times \{e, a_m\} = C_2 \times C_2 \times \cdots \times C_2,$$

and so the order of G is 2^m. This completes the proof.   ∎

Problem 7.5.1 Consider the the set of residues $\{1, 2, 3, 4\}$ mod 5. (See Section 2.2.) Verify that

$$2^2 \equiv 4 \,(\mathrm{mod}\ 5), \quad 2^3 \equiv 3 \,(\mathrm{mod}\ 5), \quad 2^4 \equiv 1 \,(\mathrm{mod}\ 5),$$

and hence verify that this set, with multiplication mod 5, is a group that is isomorphic to C_4.

Problem 7.5.2 Let G and H be groups and let P be the set of all elements of the form gh, where g and h belong to G and H, respectively. Define a binary operation that combines elements in P as follows:

$$(g_1 h_1)(g_2 h_2) = g_3 h_3,$$

where $g_3 = g_1 g_2$ is an element of G and $h_3 = h_1 h_2$ is an element of H, and where every element of G commutes with every element of H. Verify that P is a group, with identity element $e = e_G e_H$, where e_G and e_H are the identity elements of G and H, respectively, and gh has inverse element $g^{-1} h^{-1}$. Show that the group P is isomorphic to $G \times H$.

Problem 7.5.3 We have

$$C_3 \times C_3 = \{e, a, a^2\} \times \{e, b, b^2\}, \quad \text{where} \quad a^3 = b^3 = e$$

and e is the identity element of C_3. Thus the nine elements of $C_3 \times C_3$ are of the form $a^i b^j$, where $0 \le i, j \le 2$, and the elements a and b commute. Construct the group multiplication table for the abelian group $C_3 \times C_3$ and write down its proper subgroups.

7.6   Groups of Small Order

This section is devoted to finding *all* groups of order n, for $2 \leq n \leq 8$. We see from Theorem 7.3.6 that for $n = 2$, 3, 5, and 7 the only group is the cyclic group of that order. Now let us consider groups of order 4. Any group of order 4 consists of a set of the form $\{e, a, b, c\}$. If this is not the cyclic group C_4, then a, b, and c must all be of order 2. You should now check that neither ab nor ba can be equal to e, a, or b. For example, if ab were equal to e, this would imply that $b = a^{-1}$, which is impossible, since $a^{-1} = a$ and the two elements a and b are distinct. Thus we must have

$$c = ab = ba,$$

which gives the 4-group. (See Table 7.3.) Therefore there are only two groups of order 4, the cyclic group $C_4 = \{e, a, a^2, a^3\}$, and the 4-group.

Consider any group G of order 6. If G is not the cyclic group C_6, it must have an element of order 2 or 3, and since 6 is not a power of 2, it follows from Theorem 7.5.2 that G must have at least one element, say a, of order 3. Then e, a, and a^2 are distinct elements in G. Let b denote another distinct element of G, which must be of order 2 or 3. Consider the six elements

$$e, \quad a, \quad a^2, \quad b, \quad ab, \quad a^2 b. \tag{7.31}$$

We see that $ab \neq a^2 b$, for otherwise, we could deduce from the cancellation properties (see Theorem 7.2.1) that $a = e$. With a little more work, we can check that neither ab nor $a^2 b$ is equal to e, a, or a^2. Thus the six elements given in (7.31) are distinct, and this argument shows that the elements of any group of order 6, other than C_6, can be expressed in this form (with $a^3 = e$). It remains only to determine what restrictions need to be imposed on a and b to ensure that the set of elements in (7.31) forms a group. Having encountered the group of the equilateral triangle and others isomorphic to it, we know that there is at least one group of this kind.

Let us consider the closure axiom. Since $a^3 = e$, premultiplication of all six elements in (7.31) by a gives us the same elements in a different order. Let us now consider postmultiplication of all six elements by b. In particular, let us consider b^2. You should check, using the cancellation properties, that b^2 cannot be equal to b, ab, or $a^2 b$, and therefore one of the following three possibilities must hold:

$$b^2 = e, \quad b^2 = a, \quad b^2 = a^2. \tag{7.32}$$

If $b^2 = a$ or $b^2 = a^2$, then b is obviously not of order 2 and so must be of order 3. Then if we postmultiply the second and third equations in (7.32) by b, we obtain $e = ab$ and $e = a^2 b$, which are both false. The only remaining possibility in (7.32) is that

$$b^2 = e.$$

Now let us consider the element ba, which must be one of the six elements in (7.31). This element cannot be equal to e, a, a^2, or b. Thus we have

$$ba = ab \quad \text{or} \quad ba = a^2 b. \tag{7.33}$$

If $ba = ab$, we can verify that G would be abelian, and then

$$(ab)^2 = a^2 b^2 = a^2 \neq e \quad \text{and} \quad (ab)^3 = a^3 b^3 = b \neq e. \tag{7.34}$$

Thus the element ab would be of order 6, contradicting the fact that each element of G is of order 3 at most. We are therefore forced by (7.33) to accept that

$$ba = a^2 b.$$

If we premultiply the elements on each side of the latter equation by a, postmultiply them by b, and use the relations

$$a^3 = b^2 = e,$$

we obtain

$$(ab)^2 = e.$$

$\times$	e	a	a^2	b	ab	$a^2 b$
e	e	a	a^2	b	ab	$a^2 b$
a	a	a^2	e	ab	$a^2 b$	b
a^2	a^2	e	a	$a^2 b$	b	ab
b	b	$a^2 b$	ab	e	a^2	a
ab	ab	b	$a^2 b$	a	e	a^2
$a^2 b$	$a^2 b$	ab	b	a^2	a	e

TABLE 7.5. Multiplication table for the nonabelian group of order 6.

We can now write down the complete multiplication table for this group, shown in Table 7.5, by making repeated use of the relation $ba = a^2 b$. With the one-to-one correspondence

$$e \leftrightarrow e, \quad a \leftrightarrow r_1, \quad a^2 \leftrightarrow r_2, \quad b \leftrightarrow t_1, \quad ab \leftrightarrow t_2, \quad a^2 b \leftrightarrow t_3,$$

we see that this group is isomorphic to the group of the equilateral triangle (see Table 7.1). Thus there are only two groups of order 6. One is the cyclic group C_6, which is abelian. The other is the group of the equilateral triangle, which is not abelian.

Let us now consider groups of order 8. First we have the cyclic group C_8, which is abelian. Because 8 has factors 2 and 4, we immediately obtain two other abelian groups of order 8 that are formed from direct products involving the cyclic groups C_2 and C_4. These are $C_2 \times C_2 \times C_2$ and $C_4 \times C_2$, which are explored in Problems 7.6.2 and 7.6.3. For any group of order 8 other than C_8 or $C_2 \times C_2 \times C_2$, two of the three groups that we have already identified, we can assume that there is at least one element, say a, that is of order 4. Thus e, a, a^2, and a^3 are four distinct elements, and $a^4 = e$. Then if b is not one of the above four elements, the eight elements of the group are

$$e, \quad a, \quad a^2, \quad a^3, \quad b, \quad ab, \quad a^2b, \quad a^3b, \qquad (7.35)$$

for these elements are necessarily all distinct. This shows that every group of order 8, apart from C_8 and $C_2 \times C_2 \times C_2$, has elements given by (7.35).

Let us now use the closure axiom. The element b^2 must be one of the eight elements listed in (7.35). We readily see, using the cancellation properties (see Theorem 7.2.1), that b^2 cannot be equal to b, ab, a^2b, or a^3b. Then we argue that we cannot have $b^2 = a$ or $b^2 = a^3$, for these both contradict the fact that the order of b must be 2 or 4. There remain only two possibilities, that

$$b^2 = e \quad \text{or} \quad b^2 = a^2. \qquad (7.36)$$

As we will see, both of these values for b^2 lead us to groups of order 8.

First we pursue the case in which $b^2 = e$, and we consider possible values for ba. It is clear that ba must be one of the last three elements in (7.35). If $ba = ab$, we obtain an abelian group, and you should verify (see Problem 7.6.3) that it is the group $C_4 \times C_2$, already mentioned above. If $ba = a^2b$, this is equivalent to $a = b^{-1}a^2b$, and we can square this to obtain

$$a^2 = (b^{-1}a^2b)(b^{-1}a^2b) = b^{-1}a^4b = b^{-1}b = e,$$

which is impossible, since $a^2 \neq e$. Thus, when $b^2 = e$, the only possible value for ba that remains to be investigated is a^3b. If we premultiply the elements on both sides of the equation $ba = a^3b$ by a, and postmultiply by b, we obtain

$$(ab)^2 = a^4b^2 = e,$$

and so the element ab is of order 2. Then we can verify that the set whose elements are given by (7.35) and satisfy the conditions

$$a^4 = b^2 = (ab)^2 = e \qquad (7.37)$$

is the nonabelian group whose multiplication table is given in Table 7.6. It is called a *dihedral* group. We will meet a generalization of this group in the next section.

It remains only to follow up the second possibility in (7.36), that $b^2 = a^2$. In this case both a and b are of order 4. Again it is clear that ba must be one of the last three elements in (7.35).

If $ba = ab$, any resulting group will be abelian. Let us write $c = ab^3$. Then, since a and b commute, we have

$$c^2 = (ab^3)(ab^3) = a^2b^6 = a^2b^2 = a^4 = e,$$

$\times$	e	a	a^2	a^3	b	ab	a^2b	a^3b
e	e	a	a^2	a^3	b	ab	a^2b	a^3b
a	a	a^2	a^3	e	ab	a^2b	a^3b	b
a^2	a^2	a^3	e	a	a^2b	a^3b	b	ab
a^3	a^3	e	a	a^2	a^3b	b	ab	a^2b
b	b	a^3b	a^2b	ab	e	a^3	a^2	a
ab	ab	b	a^3b	a^2b	a	e	a^3	a^2
a^2b	a^2b	ab	b	a^3b	a^2	a	e	a^3
a^3b	a^3b	a^2b	ab	b	a^3	a^2	a	e

TABLE 7.6. Multiplication table for the dihedral group of order 8, in which $a^4 = b^2 = e$ and $ba = a^3b$.

and so $c = c^{-1}$ and c is of order 2. We see also that a and c commute, and

$$ac = a^2b^3 = \left(a^2b^2\right)b = a^4b = b.$$

Since $b = ac$, it is clear that all eight elements in (7.35) can be generated by the set $\{a, c\}$, where a and c are of order 4 and 2, respectively. This yields the abelian group $C_4 \times C_2$, and so gives us nothing new.

Next we have to check the possibility that $ba = a^2b$, with $b^2 = a^2$. Thus $ba = b^3$, and on multiplying on the left by b^{-1}, we would have $a = b^2 = a^2$, which is impossible. The final possibility for the case $b^2 = a^2$ is that

$$ba = a^3b,$$

and we can check that the set of eight elements given in (7.35), where a and b satisfy

$$a^4 = e, \quad a^2 = b^2, \quad ba = a^3b, \tag{7.38}$$

forms a nonabelian group. It is called the *quaternion* group, and its multiplication table is given in Table 7.7.

$\times$	e	a	a^2	a^3	b	ab	a^2b	a^3b
e	e	a	a^2	a^3	b	ab	a^2b	a^3b
a	a	a^2	a^3	e	ab	a^2b	a^3b	b
a^2	a^2	a^3	e	a	a^2b	a^3b	b	ab
a^3	a^3	e	a	a^2	a^3b	b	ab	a^2b
b	b	a^3b	a^2b	ab	a^2	a	e	a^3
ab	ab	b	a^3b	a^2b	a^3	a^2	a	e
a^2b	a^2b	ab	b	a^3b	e	a^3	a^2	a
a^3b	a^3b	a^2b	ab	b	a	e	a^3	a^2

TABLE 7.7. Multiplication table for the quaternion group, in which $a^4 = e$, $a^2 = b^2$, and $ba = a^3b$.

We saw in Example 7.2.1 that the quaternions form a group, and we found in Example 7.3.2 that the subset $\{\pm 1, \pm i, \pm j, \pm k\}$ is a subgroup of the full quaternion group. We can verify that the latter subgroup is isomorphic to the group whose multiplication table is given in Table 7.7. We can identify a in Table 7.7 with the quaternion i, and consequently we identify a^2, a^3, and $a^4 = e$ with -1, $-i$, and 1, respectively. We can identify $a^2 b$ with j, and consequently we need to make the identifications

$$b \leftrightarrow -j, \quad a^3 b \leftrightarrow k, \quad ab \leftrightarrow -k.$$

As we have seen above, there are five groups of order 8. Three of them, namely $C_2 \times C_2 \times C_2$, $C_4 \times C_2$, and C_8, are abelian, and the other two, the dihedral group and the quaternion group, are not abelian.

Example 7.6.1 By Lagrange's theorem, Theorem 7.3.4 any proper subgroup of the dihedral group of order 8 must be of order 2 or 4. We find that there are five subgroups of order 2,

$$\{e, a^2\}, \quad \{e, b\}, \quad \{e, ab\}, \quad \{e, a^2 b\}, \quad \{e, a^3 b\},$$

which are all isomorphic to C_2, and two subgroups of order 4,

$$\{e, a, a^2, a^3\} \quad \text{and} \quad \{e, a^2, b, a^2 b\}, \tag{7.39}$$

which are isomorphic to C_4 and the 4-group, respectively. We can verify that three of the above five subgroups are *normal* subgroups. These are $\{e, a^2\}$ and the two subgroups of order 4. ∎

Problem 7.6.1 Consider the abelian group $C_3 \times C_2$. Let us write

$$C_3 = (e, a, a^2) \quad \text{and} \quad C_2 = (e, b),$$

where e is the identity element in C_3 and C_2, and $a^3 = b^2 = e$. Show that

$$C_3 \times C_2 = \{e, a, a^2, b, ab, a^2 b\},$$

where $ab = ba$. With $c = a^2 b$, verify that

$$c^2 = a, \quad c^3 = b, \quad c^4 = a^2, \quad c^5 = ab, \quad c^6 = e,$$

and hence show that

$$C_3 \times C_2 = \{e, c, c^2, c^3, c^4, c^5\} = C_6.$$

Problem 7.6.2 Show that the abelian group $C_2 \times C_2 \times C_2$ can be generated by the set $S = \{a, b, c\}$, where $a^2 = b^2 = c^2 = e$, and a, b, c all commute, so that the elements of $C_2 \times C_2 \times C_2$ are

$$e, \quad a, \quad b, \quad c, \quad ab, \quad ac, \quad bc, \quad abc.$$

Problem 7.6.3 Show that the abelian group $C_4 \times C_2$ can be generated by the set $S = \{a, b\}$, where a and b commute and $a^4 = b^2 = e$, and consequently the elements of $C_4 \times C_2$ are

$$e, \quad a, \quad a^2, \quad a^3, \quad b, \quad ab, \quad a^2b, \quad a^3b.$$

7.7 Dihedral and Polyhedral Groups

In Section 7.2 (see Figure 7.1) we considered the group of the equilateral triangle. We can generalize this to obtain the group D_n, with $n \geq 3$, based on symmetry operations on the regular n-gon, the regular polygon with n edges. We define r_j, for $1 \leq j \leq n-1$, as the operation in which the regular n-gon is rotated through an angle $2j\pi/n$, and define e as the identity element. Let V_1 and V_2 denote two adjacent vertices of a regular n-gon whose center is O. Then the angle V_1OV_2 is $2\pi/n$, and we can see that the operation r_j rotates each vertex of the n-gon j places in a counterclockwise direction.

We also define n further operations, $t_1, t_2, \ldots, t_n$. When n is odd, each operation t_j is defined as a reflection of the n-gon in an axis that joins a vertex to the midpoint of the opposite edge. When n is even, $\frac{1}{2}n$ of the operations t_j are concerned with reflections in an axis joining opposite vertices of the n-gon; the remaining $\frac{1}{2}n$ operations t_j involve reflections in an axis joining the midpoints of opposite edges.

Since each operation t_j is a reflection, we see that $t_j^2 = e$, where e is the identity operation. Therefore, each t_j is its own inverse. If we apply the operation r_j followed by r_k, we obtain the operation r_jr_k, in which the n-gon is rotated through an angle $2(j+k)\pi/n$. Thus we have

$$r_jr_k = \begin{cases} e, & \text{if } j+k = n, \\ r_{j+k}, & \text{if } j+k < n, \\ r_{j+k-n}, & \text{if } j+k > n. \end{cases} \tag{7.40}$$

It is clear from (7.40) that the inverse of r_j is r_{n-j}. Since the operation t_jt_k involves two reflections, it is equivalent to turning the n-gon over twice and so must have the same effect as one of the rotation operations r_m. However, any operation of the form t_jr_k or r_jt_k is equivalent to turning the n-gon over once and rotating it, or vice versa, and so must yield one of the operations t_m. It is easy to verify that D_n is a group. It is called the *dihedral group* of order $2n$. We can show that the elements of D_n satisfy the following properties:

$$r_j^n = e, \quad (t_k)^2 = (r_jt_k)^2 = (t_kr_j)^2 = e \tag{7.41}$$

for $1 \leq j \leq n-1$, $1 \leq k \leq n$, and

$$r_j = r_1^j \quad \text{for } 1 \leq j \leq n-1. \tag{7.42}$$

It is easily verified that the simplest dihedral group, D_3, is the group of the equilateral triangle, as shown in Figure 7.1. We saw earlier that D_3 is nonabelian, and we will see that this is true for all dihedral groups D_n. The following example will give us some further insights into the structure of D_n.

Example 7.7.1 Let us use the square depicted in Figure 7.2 to illustrate the dihedral group D_4. The operation r_1 rotates the square counterclockwise through an angle $\frac{\pi}{2}$, so that V_1 is moved to V_2, V_2 is moved to V_3, and so on. Similarly, the operation r_2 rotates the square counterclockwise through an angle π, so that V_1 is moved to V_3, and r_3 rotates the square counterclockwise through an angle $\frac{3\pi}{2}$, so that V_1 is moved to V_4.

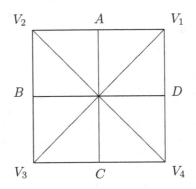

FIGURE 7.2. The dihedral group D_4.

We saw in Section 7.6 that there are only five different groups of order 8, and thus the dihedral group D_4 must be one of these. The group D_4 has two elements (r_1 and r_3) of order 4 and five elements (r_2 and the four elements t_j) of order 2, and a careful check reveals that it must be the same as the group whose multiplication table is given in Table 7.6, which I called a *dihedral* group. We begin by identifying the element r_1 in D_4 with the element a in Table 7.6. Then $r_2 = a^2$ and $r_3 = a^3$. Since we can equate the identity elements, it remains to find elements in Table 7.6 that can be identified with the reflection operations t_j. Let t_1 denote the operation of reflecting the square in the axis AC, say, and let us identify t_1 with the element b in Table 7.6. (The choice of AC is arbitrary; I could equally have chosen V_2V_4, BD, or V_1V_3.) Then let us denote r_1t_1 by t_2. This tells us that the operation t_2 corresponds to a rotation about the axis V_1V_3, and that t_2 must be identified with ab. Since $a^2b = r_1^2t_1 = r_2t_1$, we see that this operation corresponds to a rotation about the line BD in Figure 7.2. Let us denote r_2t_1 by t_3. Finally, we see that $a^3b = r_1^3t_1 = r_3t_1$, and find that this corresponds to a rotation about the line V_2V_4 in Figure 7.2. We denote

$r_3 t_1$ by t_4, and this completes the verification that the group D_4 defined in this section and the group defined in Table 7.6 are isomorphic.   ∎

Example 7.7.1 shows us that $a = r_1$ and $b = t_1$ are generators for the dihedral group D_4. We can generalize this to any dihedral group. Let us begin with D_n, whose elements are $e, r_1, \ldots, r_{n-1}$, and $t_1, \ldots, t_n$, as defined above. To help us visualize what is to follow, let us place the regular n-gon on the page so that the axis associated with one of the t_j is vertical, as we have for the square in Figure 7.2. We then write $a = r_1$, and let b denote the element t_j that is associated with the vertical axis. Then let us carry out the operations b, a, and b again. Thus we reflect the regular n-gon in the vertical axis, rotate it counterclockwise through an angle $2\pi/n$, and again reflect it in the vertical axis. You should be able to see that this is equivalent to rotating the n-gon *clockwise* through an angle $2\pi/n$. This equivalence corresponds to the algebraic statement

$$bab = a^{-1}. \tag{7.43}$$

If we multiply both sides of (7.43) on the left by $a^n = e$ and on the right by b, then, since $b^2 = e$, we obtain

$$ba = a^{n-1}b. \tag{7.44}$$

This generalizes the relations $ba = a^2b$ and $ba = a^3b$ that we obtained for $n = 3$ in Table 7.5 and $n = 4$ in Table 7.6.

Now consider the set

$$S = \left\{e, a, a^2, \ldots, a^{n-1}, b, ab, a^2b, \ldots, a^{n-1}b\right\}. \tag{7.45}$$

Since e, a, and b, all belong to D_n, it follows from the closure condition that all elements in the set S belong to D_n. We can also see, using the cancellation properties (see Theorem 7.2.1), that the $2n$ elements of S are all distinct, and so must be the elements of D_n. Since the powers of a give the elements a_j of D_n, the remaining elements of S (excepting the identity element e) must give the t_j, and so

$$\left\{b, ab, a^2b, \ldots, a^{n-1}b\right\} = \left\{t_1, t_2, \ldots, t_n\right\}.$$

Thus D_n is generated by the set $\{a, b\}$ whose elements satisfy

$$a^n = b^2 = (ab)^2 = e. \tag{7.46}$$

Note that every element of D_n can be expressed in the form $a^j b^k$, where $0 \le j \le n - 1$ and $0 \le k \le 1$. This is called a *normal* form.

There are groups that are inspired by the symmetries of the regular polyhedra. We will now consider in detail the simplest of these, the *tetrahedral group* T. You will find it almost essential to make yourself a model of the regular tetrahedron. I recommend that you label its vertices 1, 2, 3, and 4, to help you work through the following account.

Let us denote the vertices of the tetrahedron by V_1, V_2, V_3, and V_4. There are two types of symmetry operations. The first involves rotating the tetrahedron around the line joining a vertex to the centroid of the opposite face. Since there are two possible rotations corresponding to each vertex, there are eight such operations. We find that we can express these as products of cycles (see Section 7.4), and we will write

$$r_1 = (1)(234), \quad r_2 = (2)(143), \quad r_3 = (3)(124), \quad r_4 = (4)(132). \quad (7.47)$$

The other four such operations of this kind can be expressed as r_1^2, r_2^2, r_3^2, and r_4^2. We have $r_1^2 = (1)(243)$, and we can write down similar expressions for r_2^2, r_3^2, and r_4^2. Note that under the operations r_1 and r_1^2, the vertices V_2, V_3, and V_4 are moved in cyclic order.

The other kind of symmetry operation on the tetrahedron is to rotate it around the line joining the midpoints of opposite edges, and this is where you will find your model tetrahedron particularly helpful. Since the six edges of the tetrahedron can be regarded as three pairs of opposite edges, there are three such operations. We find that these can also be expressed as products of cycles, and will write them as

$$t_1 = (12)(34), \quad t_2 = (13)(24), \quad t_3 = (14)(23). \quad (7.48)$$

The eight operations r_j and r_j^2 are of order three, and the three operations t_j are of order two. Thus r_j and r_j^2 are inverses of each other, and t_j is its own inverse. We will see that the above 11 operations, together with the identity operation e, form a group of order 12. You should verify that the twelve operations, which we have expressed as permutations, are all even permutations. (See Definition 7.4.3.) In fact, since these are *all* of the even permutations on 4 objects, T is simply the alternating group A_4, which was introduced in Definition 7.4.4.

In Section 6.5 we used the notation $\{n, m\}$ to denote the regular polyhedron that is constructed from regular polygons with n sides, with m such polygons meeting around each vertex. We can find a group based on the symmetries of any regular polyhedron. However, any rotation that takes the polyhedron $\{n, m\}$ into itself also takes its dual $\{m, n\}$ into itself. Consider, for example, a regular octahedron containing an inscribed cube, whose vertices are the centers of the faces of the octahedron. Any rotation that takes the octahedron into itself will do the same for its inscribed cube. Thus the cube and octahedron have the same group, and similarly, the dodecahedron and icosahedron have the same group. These two groups are called the *octahedral* group and the *icosahedral group*. As we saw in Problem 6.5.1, an octahedron can be obtained by truncating the corners of a tetrahedron by slicing through the midpoints of its edges. The six vertices of the octahedron are the midpoints of the edges of the original tetrahedron. The ocahedron can be thought of as having two sets of four faces, one set being those that were part of the faces of the original tetrahedron, and the other being those created by the four truncations. Thus the symmetries of the octahedron *include* the symmetries of the tetrahedron, which are described by the alternating group A_4, consisting of the 12 even permutations of four objects. The only other other symmetries of the octahedron involve interchanging opposite faces. Let A denote the centroid of a face F of the octahedron that was part of the original tetrahedron, and let A' be the centroid of the face F' that is opposite to F. Thus the face F' is one of those that were created by truncating the tetrahedron. We obviously obtain three further symmetries of the octahedron by rotating it about its center so that F and F' are interchanged and then rotating it around the axis AA'. Since there are four sets of opposite faces, this yields 12 further symmetries of the octahedron. It can be shown that these correspond to the *odd* permutations of four objects. Thus the octahedral group is the full symmetric group S_4, which is of order 24. It can be shown (see Coxeter [4]) that the icosahedral group is the alternating group A_5, which is of order 60.

There is an infinite number of groups, which I will now define, whose discovery was inspired by the symmetries of the regular polyhedra.

Definition 7.7.1 The *polyhedral group* (l, m, n), for integers $l, m, n > 1$, is generated by the set $\{a, b, c\}$ whose elements satisfy

$$a^l = b^m = c^n = abc = e. \qquad \blacksquare \qquad (7.49)$$

It can be shown that this definition gives a finite group when

$$\frac{1}{l} + \frac{1}{m} + \frac{1}{n} > 1, \qquad (7.50)$$

and gives an infinite group when this inequality does not hold. It is easily verified that the only finite groups of this kind are of the form $(2, 2, n)$ with $n \geq 2$, or of the form $(2, 3, n)$ with $n = 3$, 4, or 5. Further investigation reveals that the group $(2, 2, n)$, which is expressed by three generators in (7.49), is the dihedral group D_n, which we expressed in terms of two generators in (7.46). It can also be verified that $(2, 3, 3)$ gives the tetrahedral group A_4, which we discussed above, and that $(2, 3, 4)$ and $(2, 3, 5)$ are the octahedral group S_4 and the icosahedral group A_5, respectively. For further material on groups, see, for example, P. M. Cohn [2].

Problem 7.7.1 Let a and b denote generators of the dihedral group D_n defined by (7.46). Show by mathematical induction that $ba^j = a^{n-j}b$ for $1 \leq j \leq n - 1$, and thus show how the product of any two elements of D_n can be expressed in normal form.

Problem 7.7.2 With the notation as defined in Problem 7.7.1, verify that

$$\left(a^j b\right) a^k = a^{n+j-k} b \quad \text{and} \quad \left(a^j b\right) \left(a^k b\right) = a^{n+j-k},$$

where the exponent $n+j-k$ can be replaced by $j-k$ if $j \geq k$. Hence write down a multiplication table for D_5.

Problem 7.7.3 Find all the proper subgroups of the dihedral group D_n.

Problem 7.7.4 Find all the proper subgroups of the tetrahedral group.

Problem 7.7.5 Verify that $\{e, t_1, t_2, t_3\}$, where the elements t_j are defined in (7.48), is the Klein 4-group and is a normal subgroup of the tetrahedral group.

Problem 7.7.6 Let r_1 and t_1 denote the elements of the tetrahedral group defined in (7.47) and (7.48). Let $a = r_1$ and $b = t_1$. Verify that

$$a^3 = b^2 = (ab)^3 = e.$$

Show that the set $\{a, b\}$ generates the tetrahedral group.

Problem 7.7.7 Let $u = r_1$ and $v = r_1^2 t_1$, where r_1 and t_1 are the elements of the tetrahedral group defined in (7.47) and (7.48). Verify that

$$u^3 = v^3 = (uv)^2 = e,$$

and that u and v are generators for the tetrahedral group.

Problem 7.7.8 Verify that the only solutions (l, m, n) of the inequality (7.50), where $l, m, n > 1$ are integers, are of the form $(2, 2, n)$ with $n \geq 2$, or of the form $(2, 3, n)$ with $n = 3$, 4, or 5.

References

[1] Arthur T. Benjamin and Jennifer J. Quinn. *Proofs that really count. The art of combinatorial proof*, The Dolciani Mathematical Expositions, 27, Mathematical Association of America, Washington, DC, 2003.

[2] P. M. Cohn. *Algebra Volume 1, Second Edition*, John Wiley and Sons, Chichester, 1982.

[3] John H. Conway and Richard K. Guy. *The Book of Numbers*, Springer-Verlag, New York, 1996.

[4] H. S. M. Coxeter. *Introduction to Geometry*, John Wiley & Sons, New York, 1961.

[5] P. J. Davis. The Rise, Fall, and Possible Transfiguration of Triangle Geometry: A Mini-history, *American Mathematical Monthly XX*, 204–214, 1995.

[6] Marcus du Sautoy. *The Music of the Primes*, Harper Collins, 2003.

[7] C. H. Edwards, Jr. *The Historical Development of the Calculus*, Springer-Verlag, New York, 1979.

[8] Hans Magnus Enzensberger. *Drawbridge Up: Mathematics – A Cultural Anathema* (English translation of *Zugbrücke ausser Betrieb: Die Mathematik im Jenseits der Kultur*), A. K. Peters, Natick, Massachusetts, 1999.

[9] H. Eves. *An Introduction to the History of Mathematics, 5th Edition*, Saunders, Philadelphia, 1983.

[10] Ralph P. Grimaldi. *Discrete and Combinatorial Mathematics: an Applied Introduction, 3rd Edition*, Addison-Wesley, Reading, Massachusetts, 1994.

[11] G. H. Hardy. *A Mathematician's Apology*, Cambridge University Press, 1940. Reprinted with Foreword by C. P. Snow, 1967.

[12] G. H. Hardy and E. M. Wright. *An Introduction to the Theory of Numbers, 5th Edition*, Clarendon Press, Oxford, 1979.

[13] Paul Hoffman. *The Man Who Loved Only Numbers: The Story of Paul Erdős and the Search for Mathematical Truth*, Fourth Estate, London, 1998.

[14] George M. Phillips. *Two Millennia of Mathematics: From Archimedes to Gauss*, Springer-Verlag, New York, 2000.

[15] E. C. Titchmarsh. *The Theory of Functions, 2nd Edition*, Oxford University Press, London, 1939.

[16] Simon Singh. *Fermat's Last Theorem: The Story of a Riddle that Confounded the World's Greatest Minds for 358 Years*, Fourth Estate, London, 1997.

[17] David Wells. *The Penguin Dictionary of Curious and Interesting Geometry*, Penguin Books, London, 1991.

Index